Calculus of Thought: Neuromorphic Logistic Regression in Cognitive Machines

Calculus of Thought: Neuromorphic Logistic Regression in Cognitive Machines

DANIEL M. RICE

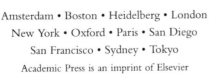

Amsterdam • Boston • Heidelberg • London
New York • Oxford • Paris • San Diego
San Francisco • Sydney • Tokyo
Academic Press is an imprint of Elsevier

ELSEVIER

Academic Press is an imprint of Elsevier
225, Wyman Street, Waltham, MA 02451, USA
The Boulevard, Langford Lane, Kidlington, Oxford OX5 1GB, UK
Radarweg 29, PO Box 211, 1000 AE Amsterdam, The Netherlands

Library of Congress Cataloging-in-Publication Data
Application submitted

British Library Cataloguing in Publication Data
A catalogue record for this book is available from the British Library

ISBN: 978-0-12-410407-5

For information on all Academic Press publications
visit our web site at store.elsevier.com

Printed and bound in USA
14 15 16 17   10 9 8 7 6 5 4 3 2 1

Working together
to grow libraries in
developing countries

www.elsevier.com • www.bookaid.org

To the memory of my father Richard Rice

CONTENTS

A PERSONAL PERSPECTIVE

I first heard the phrase Calculus of Thought as a second year graduate student in 1981 when I took a seminar called *Brain and Behavior* taught by our professor James Davis at the University of New Hampshire. He gave us a thorough grounding in the cognitive neuroscience of that day, and he spent significant time on various models of neural computation. The goal of cognitive research in neuroscience, he constantly reminded us, was to discover the neural calculus, which he took to be a complete and unifying understanding of how the brain performs computation in diverse functions such as coordinated motor action, perception, learning and memory, decision making, and causal reasoning. He suggested that if we had such a calculus, then there would be truly amazing practical artificial intelligence applications beyond our wildest dreams. This was before the widespread popularity of artificial neural network methods in the mid-1980s, as none of the quantitative models that we learned about were artificial neural networks, but instead were grounded in empirical data in neural systems like the primary visual cortex and the basal ganglia. He admitted that the models that we were taught fell way short of such a calculus, but he did fuel an idea within me that has stayed for more than 30 years.

I went on to do my doctoral dissertation with my thesis advisor Earl Hagstrom on timing relationships in auditory attention processing as reflected by the scalp recorded Electroencephalography (EEG) alpha rhythm. During the early and mid-1980s, there was not a lot of interest in the EEG as a window into normal human cognition, as the prevailing sentiment was that the brain's electric field potentials were too gross of a measure to contain valuable information. We now have abundant evidence that brain field potential measures, such as the EEG alpha rhythm, are sensitive to oscillating and synchronous neural signals that do reflect the rhythm and time course of cognitive processing. My dissertation study was one of a handful of initial studies to document a parallelism between the oscillating neural synchrony as measured by the EEG and cognition.[1] Through that dissertation study, and the advice that I also got from other faculty committee members—John Limber and Rebecca Warner, I learned the importance of well-controlled experimental findings in providing more reliable explanations. Yet, this dissertation did not address how basic cognitive computations might be performed by oscillating neural synchrony mechanisms, and my thoughts have wandered back to how this Calculus of Thought might work ever since.

I did postdoctoral fellowship research with Monte Buchsbaum at the University of California-Irvine Brain Imaging Center. Much of our work focused on abnormal temporal lobe EEG slowing in probable Alzheimer's patients and related temporal lobe slowing in nondemented older adults. We published one paper in 1990 that replicated other studies including one from Monte's lab showing abnormal temporal lobe EEG slowing in Alzheimer's patients compared to nondemented elderly, but we critically refined the methodology to get a higher fidelity measurement.[2] We published a second paper in 1991 that used this refined methodology to make the first claim that Alzheimer's disease must have an average preclinical period of at least 10 years.[3] Our basic finding was that a milder form of the temporal lobe EEG slowing observed in Alzheimer's patients was seen in nondemented older adults with minor memory impairment. This observed memory impairment had the exact same profile as what neuropsychologist Brenda Milner had observed in the famous medial temporal lobe patient H.M. This is the most famous clinical case study in neuroscience.[4] This was that there was normal immediate recall ability, but dramatically greater forgetting a short time later. Other than this memory deficit, we could find nothing else wrong in terms of cognitive and intelligence tests with these nondemented older adults. Given the nature of the memory deficit and the location of the EEG abnormality, we suggested that the focus of this abnormality must be in the medial temporal lobe of the brain. Given the prevalence of such EEG slowing in nondemented older adults and the prevalence of Alzheimer's disease, we calculated that this must be a preclinical sign of Alzheimer's disease that is present 10 years earlier than the clinical diagnosis. We had no idea how abnormal neural synchrony might be related to dysfunction in memory computations, but ever since my thoughts have wandered back to how this Calculus of Thought could go awry early in Alzheimer's disease.

Our claim that there is a long preclinical period in Alzheimer's disease with a major focal abnormality originating in the medial temporal lobe memory system has now become the prevailing view in Alzheimer's research.[5,6,7] This evidence for the long preclinical period is such that the National Institute on Aging issued new diagnostic recommendations in

2011 to incorporate preclinical Alzheimer's disease into the clinical Alzheimer's disease diagnosis.[8] Yet, when I moved to a new assistant professor position at the University of Southern California (USC) in the earlier 1990s and tried to get National Institutes of Health (NIH) funding to validate this hypothesis with a longitudinal, prospective study, my proposals were repeatedly not funded. It was not controversial that Alzheimer's starts with medial temporal lobe memory system problems, as almost everyone believed that was true. However, I was a complete newcomer, and the idea that Alzheimer's had such a long preclinical period was just too unexpected to be accepted as reasonable by the quorum needed to get NIH funding. It was then that I began to have recollections of Thomas Kuhn's theory about bias in science taught by my undergraduate history of science professor Charles Bonwell many years earlier. Kuhn warned that the established scientific community is often overinvested in the basic theory in their field that he called "paradigm".[9] Kuhn argued that this bias precludes most scientists from considering new results and hypotheses that are inconsistent with the paradigm. In the early 1990s, the entire Alzheimer's clinical diagnostic process and pharmaceutical research were only concerned with actual demented patients. So any results or explanation that there is in fact a long 10-year preclinical period was not yet welcome. As it turns out, the only hope to prevent and treat Alzheimer's disease now seems to be in this very early period when the disease is still mild,[10] as all drug studies in actual demented patients have not really worked. At that time, I was not yet familiar with Leibniz's concept of *Calculus Ratiocinator*, or a thought calculus machine designed to generate explanations that avoid such bias. Though, I did start to think that any unbiased machine that could generate most reasonable hypotheses based upon all available data would be useful.

Fortunately, I had wider research interests than Alzheimer's disease, as I had an interest in quantitative modeling of the brain's electric activity as a means to understand the brain's computations. One day while still at UC-Irvine, I attended a seminar given by a graduate student on maximum entropy and information theory in a group organized by mathematical cognitive psychologists Duncan Luce and Bill Batchelder. I then began to study maximum entropy on my own and became interested in the possibility that this could be the basic computational process within neurons and the brain. When I ended up teaching at the USC a few years later, I was fortunate enough to collaborate with engineering professor Manbir Singh and his graduate student Deepak Khosla on modeling the EEG with the maximum entropy method.[11] In our maximum entropy modeling, Deepak taught me a very interesting new way to smooth error out of EEG models using what we now call L2 norm regularization. Yet, I also began to think that there might be a better approach based upon probability theory to model and ultimately reduce error in regression models that model the brain's neural computation. This thinking eventually led to the reduced error logistic regression (RELR) method that is the proposed Calculus of Thought, which is the subject of this book.

In April of 1992, I had a bird's eye view of the Los Angeles (LA) riots through my third floor laboratory windows at USC that faced the south central section of the city. I watched shops and houses burn, and I was shocked by the magnitude of the violence. But, I also began to wonder whether human social behavior also might be determined probabilistically in ways similar to how causal mechanisms determine cognitive neural processes like attention and memory so that it might be possible to predict and explain such behavior. After the riots, I listened to the heated debates about causal forces involved in the 1992 LA riots, and again I began to wonder how objective these hypotheses about causal explanations of human behavior ever could be due to extremely strong biases. This was also true of most explanations of human behavior that I saw offered in social science whether they were conservative or liberal. So, it became clear to me that bias was the most significant problem in social science predictions and explanations of human behavior. And, I began to believe that an unbiased machine learning methodology would be a huge benefit to the understanding of human behavior outcomes. However, I did not yet make the connection that a data-driven quantitative methodology that models neural computations could be the basis of this unbiased *Calculus Ratiocinator* machine.

Eventually, it became clear that my proposed research on a longitudinal, prospective study to test the putative long preclinical period in Alzheimer's disease would not be funded. So, because NIH funding was required to get tenure at USC, I realized that I had better find another career. I resigned from USC in 1995, which was years prior to when my tenure decision would have been made. Even though I had a contract that would be renewed automatically until at least 1999, I saw no reason to be dead weight for several years and then face a negative tenure decision. Instead, I wanted to get started on a more entrepreneurial career where I would have more control over my fate. So, I pursued an applied analytics career. I have enjoyed this career often better than my academic career because the rewards and feedback are more immediate. Also, there is much more emphasis on problem solving as the goal rather than writing a paper or getting a grant, although sales ability is obviously always valued. Yet, I eventually realized that the same bias problems that limit understanding of human behavior in academic neural, behavioral, and social science also limit practical business analytics. Time and time again I noticed decisions based upon biased predictive or explanatory models dictated by incorrect, questionable assumptions, so again I noticed a need for a data-driven *Calculus Ratiocinator* to generate reasonable and testable hypotheses and conclusions that avoid human bias.

Much of my applied analytic work has involved logistic regression, and I eventually learned that maximum entropy and maximum likelihood methods yield identical results in modeling the kind of binary signals that both neurons and business decisions produce. Thus, at some point I also realized that I could continue my research into the RELR Calculus of Thought. But instead of focusing on the brain's computations, I could focus on practical real world business applications using the same computational method that I believed that neurons use—the RELR method. In so doing, I eventually realized that RELR could be useful as the *Calculus Ratiocinator* that Leibniz had suggested is needed to remove biased answers to important, real world questions.

Many of today's "data scientists" have a background in statistics, pure mathematics, computer science, physics, engineering, and operations research. Yet, these are academic areas that are not focused on studying human behavior, but instead focus on quantitative and technical issues related to more mechanical data processes. Many other analytic scientists, along with many analytic executives, have a background in behaviorally oriented fields like economics, marketing, business, psychology, and sociology. These academic areas do focus on studying human behavior, but are not heavily quantitatively oriented. There is a real need for a theoretical bridge between the more quantitative and more behavioral knowledge areas, and that is the intention of this book. So, I believe that this book could appeal both to quantitative and behaviorally oriented analytic professionals and executives and yet fill important knowledge gaps in each case.

Unless an analytic professional today has a specific background in cognitive or computational neuroscience, it is unlikely that they will have a very good understanding of neuroscience and how the brain may compute cognition and behavior. Yet, data-driven modeling that seeks to solve the Turing problem and mimic the performance of the human brain without its propensity for bias and error is probably the implicit goal of all machine-learning applications today. In fact, this book argues that cognitive machines will need to be neuromorphic, or based upon neuroscience, in order to simulate aspects of human cognition. So, this book covers the most fundamental and important concepts in modern cognitive neuroscience including neural dynamics, implicit and explicit learning, neural synchrony, Hebbian spike-timing dependent plasticity, and neural Darwinism. Although this book presents a unified view of these fundamental neuroscience concepts in relationship to RELR computation mechanisms, it also should serve as a much needed introductory overview of these fundamental cognitive neuroscience principles for analytic professionals who are interested in neuromorphic cognitive machines.

Theories in science can simplify a subject and make that subject more understandable and applicable in practice, even when theories ultimately are modified with new data. So this book may help analytic professionals understand fundamental neural computational theory and cognitive neuroscientists understand practical issues surrounding real world machine learning applications. It is true that the theory of the brain's computation in this book should be viewed as a question rather than an answer. But, if cognitive neuroscientists and analytic professionals all start asking questions about whether the brain and machine learning each can work according to the information theory principles that are laid out in this book, then this book will have served its purpose.

Daniel M. Rice
St Louis, MO (USA)
Autumn, 2013

CHAPTER 1

Calculus Ratiocinator

"It is obvious that if we could find characters or signs suited for expressing all our thoughts as clearly and as exactly as arithmetic expresses numbers or geometry expresses lines, we could do in all matters insofar as they are subject to reasoning all that we can do in arithmetic and geometry. For all investigations which depend on reasoning would be carried out by transposing these characters and by a species of calculus."

Gottfried Leibniz, Preface to the General Science, 1677.[1]

Contents

At the of end of his life and starting in 1703 Gottfried Leibniz engaged in a 12-year feud with Isaac Newton over who first invented the calculus and who committed plagiarism. All serious scholarship now indicates that both Newton and Leibniz developed calculus independently.[2] Yet, stories about Leibniz's invention of calculus usually focus on this priority dispute with Newton and give much less attention to how Leibniz's vision of calculus differed substantially from Newton's. Whereas Newton was trained in mathematical physics and continued to be associated with academia during the most creative time in his career, Leibniz's early academic failings in math led him to become a lawyer by training and an entrepreneur by profession.[3] So Leibniz's deep mathematical insights that led to calculus occurred away from a university professional association. Unlike Newton whose entire mathematical interests seemed tied to physics, Leibniz clearly had a much broader goal for calculus. These applications were in areas well beyond physics that seem to have nothing to do with mathematics. His dream application was for a *Calculus Ratiocinator*, which is synonymous with Calculus of Thought.[4] This can be interpreted to be a very precise mathematical model of cognition that could be automated in a machine to answer any important philosophical, scientific, or practical question that traditionally would be answered with human subjective conjecture.[5] Leibniz proposed that if we had such a cognitive calculus, we could just say "Let us calculate"[6] and always find most reasonable answers uncontaminated by human bias.

In a sense, this concept of *Calculus Ratiocinator* foreshadows today's predictive analytic technology.[7] Predictive analytics are widely used today to generate better than chance longer term projections for more stable physical and biological outcomes like climate change, schizophrenia, Parkinson's disease, Alzheimer's disease, diabetes, cancer, optimal crop yields, and even good short-term projections for less stable social outcomes like marriage satisfaction, divorce, successful parenting, crime, successful businesses, satisfied customers, great employees, successful ad campaigns, stock price changes, loan decisions, among many others. Until the widespread practice of predictive analytics with the introduction of the computers in the past century, most of these outcomes were thought to be too capricious to have anything to do with mathematics. Instead, they were traditionally answered with speculative and biased hypotheses or intuitions often rooted in culture or philosophy (Fig. 1.1).

Figure 1.1 Gottfried Wilhelm Leibniz.[8]

Until just very recently, standard computer technology could only evaluate a small number of predictive features and observations. But, we are now in an era of big data and high performance massively parallel computing. So our predictive models should now become much more powerful. This is because it would seem reasonable that those traditional methods that worked to select important predictive features from small data will scale to high-dimension data and suddenly select predictive models that are much more accurate and insightful. This would give us a new and much more powerful big data machine intelligence technology that is everything that Leibniz imagined in a *Calculus Ratiocinator*. Big data massively parallel technology should thus theoretically allow completely new data-driven cognitive machines to predict and explain capricious outcomes in science, medicine, business, and government.

Unfortunately, it is not this simple. This is because observation samples are still fairly small in most of today's predictive analytic applications. One reason is that most real-world data are not representative samples of the population to which one wishes to generalize. For example, the people who visit Facebook or search on Google might not be a good representative sample of many populations, so smaller representative samples will need to be taken if the analytics are to generalize very well. Another problem is that many real-world data are not independent observations and instead are often repeated observations from the same individuals. For this reason, data also need to be down sampled significantly to be independent observations. Still, another problem is that even when there are many millions of independent representative observations, there are usually a much smaller number of individuals who do things like respond to a particular type of cancer drug or commit fraud or respond to an advertising promotion in the recent past. The informative sample for a predictive model is the group of targeted individuals and a group of similar size that did not show such a response, but these are not usually big data samples in terms of large numbers of obser-vations. So, the biggest limitation of big data in the sense of a large number of observations is that most real-world data are not "big" and instead have limited numbers of observations. This is especially true because most pre-dictive models are not built from Facebook or Google data.[9]

Still, most real-world data are "big" in another sense. This is in the sense of being very high dimensional given that interactions between variables and nonlinear effects are also predictive features. Previously we have not had the technology to evaluate high dimensions of potentially predictive variables rapidly enough to be useful. The slower processing that was the reason for this "curse of dimensionality" is now behind us. So many might believe that this suddenly allows the evaluation of almost unfathomably high dimensions of data for the selection of important features in much more accurate and smarter big data predictive models simply by applying traditional widely used methods.

Unfortunately, the traditional widely used methods often do not give unbiased or non-arbitrary predictions and explanations, and this problem will become ever more apparent with today's high-dimension data.

1. A FUNDAMENTAL PROBLEM WITH THE WIDELY USED METHODS

There is one glaring problem with today's widely used predictive analytic methods that stands in the way of our new data-driven science. This problem is inconsistent with Leibniz's idea of an automated machine that can reproduce the very computations of human cognition, but without the subjective biases of humans. This problem is suggested by the fact that there are probably at least hundreds of predictive analytic methods that are in use today. Each method makes differing assumptions that would not be agreed upon by all, and all have at least one and sometimes many arbitrary parameters. This arbitrary diversity is defended by those who believe a "no free lunch" theorem that argues that there is no one best method across all situations.[10,11] Yet, when predictive modelers test various arbitrary algorithms based upon these methods to get a best model for a specific situation, they obviously will only test but a tiny subset of the possibilities. So unless there is an obvious very simple best model, different modelers will almost always produce substantially different arbitrary models with the same data.

As examples of this problem of arbitrary methods, there are different types of decision tree methods like CHAID and CART which have different statistical tests to determine branching. Even with the very same method, different user-provided parameters for splitting the branches of the tree will often give quite different decision trees that will generate very different predictions and explanations. Likewise, there are many widely used regression variable selection methods like stepwise and LASSO logistic regression that are all different in the arbitrary assumptions and parameters employed in how one selects important "explanatory" variables. Even with the very same regression method, different user choices in these parameters will almost always generate widely differing explanations and often substantially differing predictions. There are other methods like Principal Component Analysis (PCA), Variable Clustering and Factor Analysis that attempt to avoid the variable selection problem by greatly reducing the dimensionality of the variables. These methods work well when the data match underlying assumptions, but most behavioral data will not be easily modeled with the assumptions in these methods like orthogonal components in the case of PCA or that one knows how to rotate the components to be nonorthogonal using the other methods given that there are an infinite number of possible rotations. Likewise, there are many other methods like Bayesian Networks, Partial Least Squares, and Structural Equation Modeling that modelers often use to make explanatory inferences. These methods each make differing arbitrary assumptions that often generate wide diversity in explanations and predictions. Likewise, there are a large number of fairly black box methods like Support Vector Machines, Artificial Neural Networks, Random Forests, Stochastic Gradient Boosting, and various Genetic Algorithms that are not completely transparent in their explanations of how the predictions are formed, although some measure of variable importance often can be obtained. These methods can generate quite different predictions and important variables simply because of differing assumptions across the methods or differing user-defined modeling parameters within the methods.

Because there are so many methods and because all require unsubstantiated modeling assumptions along with arbitrary user-defined parameters, if you gave exactly the same data to a 100 different predictive modelers, you would likely get a 100 completely different models unless it was a simple solution. These differing models often would make very different predictions and almost always generate different explanations to the extent that the method produces transparent models that could be interpreted. In cases where regression methods are used and raw interaction or nonlinear effects are parsimoniously selected without accompanying main effects, the model's predictions are even likely to depend on how variables are scaled so that currency in Dollars versus Euros would give different predictions.[12] Because of such variability that even can defy basic principles of logic, it is unreasonable to interpret any of these arbitrary models as reflecting a causal and/or most probable explanation or prediction.

Because the widely used methods yield arbitrary and even illogical models in many cases, hardly can we say "Let us calculate" to answer important questions such as the most likely contribution of environmental versus genetic versus other biological factors in causing Parkinson's disease, Alzheimer's disease, prostate cancer, breast cancer and so on. Hardly, can we say "Let us calculate", when we wish to provide a most likely explanation for why there is climate change or why certain genetic and environmental markers correlate to diseases or why our business is suddenly losing customers or how we may decrease costs and yet improve quality in health care. Hardly, can we say "Let us calculate", when we wish to know the extent to which sexual orientation and other average gender differences are determined by biological factors or by social factors, when we wish to know whether stricter guns control policies would have a positive or negative impact on crime and murder rates, or when we wish to know whether austerity as an economic intervention tool is helpful or hurtful. Because our widely used predictive analytic methods are so influenced by completely subjective human choices, predictive model explanations and predictions about human diseases, climate change, and business and social outcomes will have substantial variability simply due to our cognitive biases and/or our arbitrary modeling methods. The most important questions of our day relate to various economic, social, medical, and environmental outcomes related to human behavior by cause or effect, but our widely used predictive analytic methods cannot answer these questions reliably.

Even when the very same method is used to select variables, the important variables that the model selects as the basis of explanation are likely to vary across independent observation samples. This sampling variability will be especially prevalent if the observations available to train the model are limited or if there are many possible features that are candidates for explanatory variables, and if there is also more than a modest correlation between at least some of the candidate explanatory variables. This problem of correlation between variables or *multicollinearity* is ultimately the real culprit. This multicollinearity problem is almost always seen with human behavior outcomes. Unlike many physical phenomena, behavioral outcomes usually cannot be understood in terms of easy to separate uncorrelated causal components. Models based upon randomized controlled experimental selection methods can avoid this multicollinearity problem through designs that yield variables that are orthogonal.[13] Yet, most of today's predictive analytic applications necessarily must deal with observation data, as randomized experiments are simply not possible usually with human behavior in real-world situations. Leo Breiman, who was one of the more prominent statisticians of recent memory, referred to this inability to deal with multicollinearity error as "the quiet scandal of statistics" because the attempts to avoid it in traditional predictive modeling methods are arbitrary and problematic.[14]

There is a wonderful demonstration of this Breiman "quiet scandal" problem in a paper by Austin and Tu in 2004.[15] They were interested in the possibility of using binary logistic regression to answer "Why do some people die after a heart attack?". With only 29 candidate variables, they observed 940 models with differing selected variables out of 1000 total models based upon different training samples. They found similar instability no matter the variable selection methods—backward, forward or stepwise variable selection. To further the problem beyond just this sampling issue, each of these variable selection methods has their own set of arbitrary parameters. These include the level of significance for a variable to be selected, whether and how correlated variables are removed initially before any variable selection and whether further criteria are employed to ensure a parsimonious model.[16] So the amount of variability in these variable selection models produced from just 29 variables in this data set must be truly enormous. What compounds this problem even further is that the regression weight corresponding to a given selected variable often varies dramatically and even reverses in sign depending on other selected variables in a solution.[17]

So not only do we have a problem where different widely used methods generate very different models but even the same widely used variable selection method is likely to generate quite different models on separate occasions simply because of multicollinearity error and/or incorrect parameter choices. Because observations are usually limited, multicollinearity error will be a problem with these widely used predictive methods in most applications. Cognitive bias in the selection of models also will be a problem even with large numbers of observations. Unreliability due to multicollinearity error and cognitive bias is even seen with less than 30 candidate features like in the Austin and Tu study, where expert judgments to override the automated stepwise methods also produced highly unreliable variable selections.[18] In today's Big Data high-dimension problems where potentially millions and even billions of candidate features are possible when interaction and nonlinear effects are derived from a large number of input variables, the extent of variability simply due to multicollinearity error and cognitive bias is almost astronomical.

This variability is usually explained in the predictive analytics community with the claim that building a model is much more of an art than a science. It is true that the arbitrary parameters and assumptions within the widely used methods allow the modeler to have substantial artistic license to choose the final form of the model. For this reason, a situation where each modeler tells a different explanatory story and makes different predictions with the same data is definitely much more like art than science. But, this actually is the heart of the problem because we do not wish engage in art, but instead hope to have a science. So, if we seek a mechanical *Calculus Ratiocinator* that is completely devoid of cognitive bias and multicollinearity error in the prediction and explanation of outcomes, today's widely used predictive analytic methods will not get us there. In fact, these methods will likely cause greater confusion than resolution because they will all tend to give such entirely different solutions based upon the method employed, the arbitrary and subjective choices of the person who built the model, and the data sample used.

While diverse and arbitrary explanations are the most glaring problem, diverse and arbitrary predictions can be the most risky problem especially with multicollinear high-dimension data. With high-dimension genetic microarray data, the MAQC-II study of 30,000 predictive models reported substantial evidence of predictive variability across different modelers, data samples, and different traditional variable selection methods.[19] Because this predictive modeling variability directly causes decision making variability, this problem creates a substantial amount of risk and uncertainty. In fact, this MAQC-II study reported that the single best predictor of model failure in the sense of poor blinded sample predictive accuracy was when models built from the very same data set had discrepant variable selection across different modelers and methods. The MAQC-II study suggested that this was likely to occur unless it was a rare and very easy problem where a variable selection based upon just a few features gave a highly predictive solution. Thus, variable selection variability due to cognitive bias and/or multicollinearity error not only is indicative of a problematic explanation but also may warn of risky real-world predictions.

To reiterate, predictive modeling with observational data will be likely to generate highly suspect explanations and predictions when models are built with widely used methods due to cognitive bias and multicollinearity error problems. These problems will be even more apparent with today's high-dimension data, because there are simply many more degrees of freedom to drive even much greater variability. When we produce a model for questions like "why are we losing customers, and can we predict this loss the next month?" or "why do some people die immediately after a heart attack, and can we predict this occurrence?" or "why do some people default on their loans, and can we predict who is most at risk in a potentially changing economy?" or "why has climate change occurred, and can we stop this from occurring?" or "what is the optimal treatment regimen in terms of cost that does not sacrifice quality of health care", we would hope that the model will not change when we build the model using a different representative training data sample or another expert modeler or another method or even another measurement scale for the units of variables. Such, indeed, is likely not to be the case with our widely used methods. Unfortunately, this variability will be even greatly magnified in today's Big Data environment with ultrahigh-dimension data.

2. ENSEMBLE MODELS AND COGNITIVE PROCESSING IN PLAYING JEOPARDY

One way to avoid the risks due to an arbitrary unstable model is simply to average different elementary models produced by different arbitrary methods and/or different expert modelers. Given appropriate sampling from the population of possible models, the greater the number and the greater variety of elementary models that go into the average, the more likely it will be that the grand average ensemble model will be able to produce accurate and reliable predictions that will not be arbitrary. The exceptional Jeopardy game show performance by IBM's Watson exemplifies this power of ensemble modeling. In early 2011, we were all mesmerized by how Watson was able to perform better than the most capable human Jeopardy players. While we do not know the fine details of the ensemble machine learning process used in Watson, we do know that it is based upon an ensemble average of many elementary models.

Although there are many ways to construct an ensemble model, we also know that the most successful ensemble models, like that which won the Netflix prize, have been produced as a grand average "stacked" ensemble model across hundreds of elementary models.[20] In many other predictive machine learning applications, such blended or stacked ensemble models are usually the most accurate methods, as was evidenced by the best performing methods in the Heritage Health Prize contest in 2011. There are other ensemble methods like Random Forest and its related hybrid Stochastic Gradient Boosting that are still sometimes accurate and much easier to implement than the stacked ensemble method because they are stand-alone methods that do not require averaging models built from many different methods. Yet, methods based upon Random Forest still have significant bias and multicollinearity error problems,[21] so a more comprehensive ensemble averaging across a large number of different methods as is done in the stacked ensemble seems to be the more effective way to remove bias and error problems. Indeed, Seni and Elder's analysis of why ensemble methods perform well suggests that the averaging creates more stable predictions that are less susceptible to the error problems that would be especially problematic with multicollinear features.[22]

If we can build stacked ensemble models that are not arbitrary when a large enough assortment of models are sampled to produce the average, then this type of ensemble machine learning should give stable predictions that do not depend on the bias of any one individual modeler. Because stacked ensemble models also can simulate human cognition as in Watson's Jeopardy performance, then this may be a basis for Leibniz's *Calculus Ratiocinator*. In fact, Watson's performance seemed quite comparable with the best performing human Jeopardy players in all respects except he was faster and showed greater breadth of knowledge. Thus, Watson may have realized Leibniz's dream for an objective calculus of thought that is also a basis for our new data-driven cognitive machines. But even though Watson was highly successful at simulating human cognition, let us not move so fast. Let us first ask: what aspect of human cognition did Watson simulate?

While Watson's stacked ensemble models usually would be very good at retrieving names and other semantically appropriate words and phrases as required in Jeopardy, or in predicting movie ratings with accurate intuition as in Netflix, these models are not proficient at providing causal explanations for why a specific name or fact has been retrieved or why a certain prediction is intuited. Indeed, good prediction does not constitute a good causal explanation as we know from Ptolemy's extremely complicated ancient astronomical model that put the earth at the center of the universe. The complexity of stacked ensemble models is one reason that they do not work as explanatory models. This is because causal explanations that humans can understand are always parsimonious, as the conscious human brain has a significant capacity limit that does not allow more than a small set of conscious features to be maintained and readily accessed and understood. Easily understood explanations tend to chunk many elementary features into larger meaningful representations that are both very parsimonious and informative.[23]

Yet, the human brain is also capable of retrieving names or facts or correctly predicting actions without conscious causal explanation of the features that cause these outcomes. This happens when a familiar name pops into our mind in response to a memory cue, but we have no conscious recollection of how we know that this name is correct in terms of an episode in our lives or associated facts that caused us to know that it is true. This also happens all the time when we speak, as the words just seem to flow, and we do not usually need to retrieve a reason for why we have chosen any particular word consciously. As an example of how such automatic processing can be used in Jeopardy, the Jeopardy cue "Real estate owned with no claims on it is said to be free and ..." causes the word "clear" to pop into my mind. But, since I never took a course in business law, I have no understanding as to why this particular word is retrieved and feels familiar and gives the correct answer. Eventually, I notice later that a frequently running television commercial that I probably have seen at least 20 times but have always tried to ignore uses this very same phrase. We will see that ensemble modeling would be very good at this type of purely predictive rote learning process that requires no associated conscious causal explanation or even any conscious learning intention.

There are certainly other times when I play Jeopardy when I am conscious of the causal linkages that explain my memory. As an example, say that the Jeopardy category is First Ladies and the cue is "She was First Lady for 12 years and 39 days". In this case, I immediately know that Franklin Roosevelt was President for more than two four year terms, and I immediately realize that his wife was Eleanor. So, I respond with "Who was Eleanor Roosevelt?" In this particular Jeopardy example, some people may even remember an episode in their lives when they learned one or both of these two facts about Franklin and Eleanor Roosevelt. Ensemble modeling would not be able to produce this explanation based upon a few associated features that caused the recollection memory.

3. THE BRAIN'S EXPLICIT AND IMPLICIT LEARNING

We will eventually recognize the basic memory process that Watson seems to simulate. But first we need a more general overview of the brain's learning and memory processes. Like artificial intelligence and machine learning research, cognitive neuroscience is an evolving field where there are still many areas of knowledge that have yet to be clarified. Yet, there is now fairly good agreement among cognitive neuroscientists in viewing the brain's learning and memory processing along a slow *explicit learning* versus fast *implicit learning* continuum.[24]

Explicit learning is characterized as slower, conscious, intentional processing where dysfunction in the brain's medial temporal lobe structures including the hippocampus will disrupt storage and retrieval of more recent memories. *Episodic memory*, or the conscious memory for events in our lives, along with *semantic memory*, or the conscious memory for facts and concepts, are explicit processes because the new learning of these types of memory is disrupted by medial temporal lobe injuries. Implicit learning is characterized as a rapid, unconscious, automatic processing that takes place independently of the brain's medial temporal lobe structures. Implicit learning is exemplified in *procedural memory* skills such as bicycle riding, typing, walking, piano playing, and automobile driving.

Working memory can be described as a limited capacity short-term conscious memory system that is very important in learning and which interacts with and controls attention. Working memory is believed to arise through the temporary reactivation and reconfiguring of existing *long-term memory* representations.[25,26] Working memory tasks can involve either *deep processing* or *shallow processing* tasks in what might be considered intermittent cached updating of long-term memory. *Deep processing* requires effort and is slow because it elaborates on the meaning of stimuli, as when we determine whether words that we read are a member of a certain category such as "fish". The transfer of new associations into longer term explicit memories is likely to result from deep processing. Yet, long-term explicit memories are much less likely to result from *shallow processing* like when we determine whether words visually presented to us are in upper or lower case.[27] But, shallow processing does lead to unconscious implicit long-term memories due to rote repetition and *priming*. Repetition priming is when the previous presentation of a stimulus such as a word makes it more likely that this word or a semantically related word will be remembered even though we may not recall the previous presentation episode. Implicit memories due to rote repetition priming can be learned very rapidly with very few repetition trials.[28]

An important book by Nobel Laureate Daniel Kahneman titled *Thinking, Fast and Slow* reviews how these slow, deep, explicit and fast, shallow, implicit cognitive mechanisms are fundamental processes that define how the brain works and how human brain decisions are determined.[29] Intuitive snap judgments are examples of fast, shallow, implicit processing, whereas well-reasoned causal explanations are examples of slow, deep, explicit processing. Much of Kahneman's book is concerned with how subjective cognitive biases cause errors in human decisions. My task in this book will be to propose how these implicit and explicit brain learning modes can be understood and modeled in terms of underlying machine learning mechanisms. Such understanding would allow cognitive machines similar to Leibniz's concept of a *Calculus Ratiocinator* that are devoid of arbitrary subjective biases and assumptions.

Playing the piano can be a good example of implicit learning. I took piano lessons for several years when I was in my early twenties, and my piano teacher incorporated a simple reward learning feedback process into his teaching. The teacher's rule was that I had to play a very small section of a piece 10 times in a row correctly before I could move to the next section; he would then apply the same principle to larger sections that grouped together the smaller sections. In its most mechanical form, the only feedback provided was a label that the targeted small section had been played correctly or incorrectly without any reason for why. As I began to play the target piece many times in a row correctly, I would notice that my mind would begin to wander. So I no longer needed to be conscious of the individual notes or chords that I was playing or even whether or not the teacher was about to give me feedback that I had played it incorrectly. When that occurred, my brain was able to predict and retrieve appropriate movements without my conscious awareness of this processing.

A similar story could be told for any implicit learning due to rote repetition. Conscious involvement is not necessary to retrieve these implicit memories or even to learn the causal explanations as to why they are remembered, although some reinforcement signal apparently must be present during procedural learning which requires motor responses. In implicit memory learning, our behavior eventually becomes guided by feeling rather than by thinking. During implicit memory retrieval, our performance even becomes impaired when we think too much, as when a basketball player may choke at the free throw line when they think about the causal steps in successful execution.

Usually, explicit and implicit learning happen concurrently in the normal human brain, but brain injury studies afford a way to observe these processes independently. As an example, there is striking brain injury patient evidence that implicit memories can be learned and retrieved without a functioning explicit memory learning system. New York Times writer Charles Duhigg is probably best known in the predictive analytics community for popularizing the story about the American retailer company Target's efforts to predict buying habits related to pregnancy in his book *The Power of Habit*.[30] Yet, Duhigg's book also contains quite informative material about the neuroscience of implicit memory. For example, Duhigg reports on his interviews with cognitive neuroscientist Larry Squire of the University of California—San Diego about a medial temporal lobe encephalitis brain injury patient known in the medical literature as E.P. What is most striking about E.P. was that although his brain injury destroyed his medial temporal lobe's ability to form new explicit memories, he was still able to learn and retrieve new implicit memories.[31] When E.P.'s family moved to a new town after his brain injury occurred, he acquired the habit of going on regular long walks. Amazingly, he would never get lost even though he had no conscious recollection of these walks. Thus, rote repetitive inputs that are associated with routine responses that are often rewarded are apparently all that is needed to learn new complex implicit procedural memories even without any explicit memory recall. Just like piano playing is a skill that can be acquired by chunking together elementary behaviors with reinforcement learning to result in a complex procedural memory, navigating a walk through a completely novel environment also appears to be such a skill.

The work of Ann Graybiel[32] indicates that neural circuits involving the basal ganglia are necessary for implicit skill learning. This appears to arise from the transfer of neocortical input signals to basal ganglia output signals. As many as 10,000 neocortical neurons may converge on a single projection neuron in the striatum, which is the largest input component to the basal ganglia. These striatum projection neurons also converge with an estimated convergence rate of about 80:1 onto target neurons in primates. Thus, each feed-forward projection stage is similar to how a purely predictive machine ensemble model averages large numbers of elementary submodels into a single predictive output signal. Although this implicit memory computational process has more than a superficial resemblance to stacked ensemble predictive modeling, it is obviously much more complex given the large number of striatum neurons that all generate output predictions in parallel.[33]

This apparent explicit versus implicit dichotomy actually may be more of a continuum. This is suggested by the feeling of *familiarity* that occurs when we know that we have experienced an event previously, but cannot retrieve the causal reason. This familiarity process has some properties that are akin to more implicit memories in that it is automatic and rapid and does not involve conscious recollection of the reason for the memory. But, familiarity memory also has properties that are like more explicit memories because it involves the conscious awareness of familiarity and is disrupted in medial temporal lobe brain injuries involving the part of the medial temporal lobe which is adjacent to the hippocampus.[34–36] Such an intermediate process might result from partial forgetting of previously learned explicit memories, or from partial learning of explicit memories.

What kind of memory would it take for a machine to play Jeopardy very rapidly and efficiently? At a minimum this would require automatic and rapid memory mechanisms like primed implicit memories that were learned simply through rote repetition. This mechanism would respond to a cue like

"Real estate owned with no claims on it is said to be free and ..." by choosing the highest probability prediction automatically based upon previous rote repetition learning. This is consistent with how Watson's ensemble memory worked, as three most likely memories associated with each Jeopardy item were remembered by Watson who simply chose the most probable response. This type of implicit memory mechanism would be much faster than trying to recollect the causal linkages for the memory through conscious explicit semantic associations or previous learning episodes. So humans may also perform better in Jeopardy when they rely upon an automatic implicit memory mechanism based upon rote repetition priming. In fact, one of Watson's main advantages seemed to be that he was simply faster than the two human contestants.

Because machine stacked ensemble methods can simulate the implicit memory performance of humans, it may be reasonable to theorize that they are similar to the brain's implicit memory ensemble computations. Yet, the analogy is not perfect. This is because highly successful stacked ensemble machine learning is not automatic implicit learning like that which occurs in the brain. Instead, a human is required to decide which elementary models go into the stacked ensemble and to decide upon tuning parameters that weight the impact of individual models. In the most accurate stacked ensemble modeling, these manual interventions are typically trial and error processes that may take weeks, months, or even years to produce the best final model, whereas the brain can learn implicit memories very quickly in a small number of learning trials as in rote repetition priming learning. Additionally, machine ensemble models are implemented as scoring models where they no longer learn based upon new data, whereas the brain's implicit memory learning is capable of automatic continued new learning as in the case of brain injury patient E.P. Ideally, any *Calculus Ratiocinator* should be capable of automatic learning because it removes much of the potential for human bias.

Although this necessary manual involvement in the learning phase is a limitation, today's machine ensemble models are still very good at implicit memory tasks that do not require a causal explanation for their performance. Whether it is a natural language process in Jeopardy, a recommendation process in Netflix,[37] or the host of other purely predictive implicit memory processes in today's machine learning, ensemble models perform extremely well as long as the new predictive environment is stable and not different from the training environment. Yet, these accurate ensemble models are still limited in being just black box predictive machines. In contrast, any *Calculus Ratiocinator* must be able to generate causal explanations like those that arise from explicit memory learning.

In a sense, the computational models that we can now construct with artificial intelligence are an extension of our own brain. But, this extension can be a less intelligent system which is only able to generate short-term predictions with no causal understanding as might occur in insects, or it can be a very intelligent system that is also able to generate causal explanations as occurs in intelligent mammals. To understand the importance of explicit memory learning in intelligent systems, we need to remember the famous patient H.M. After his surgery that removed both medial temporal lobes, H.M. completely lost the ability to learn new causal explanations. Every time that he shifted the focus of his working memory, it was like he was constantly waking and not understanding the causal reason for how he got to his present situation. As an old man, he still believed that he was the same young man who had undergone the surgery, as he did not even recognize himself in the mirror. Obviously, we want our artificial intelligence models to learn the causal reasons for their predictions or at least to assist us in understanding these causal reasons, or else they will not be very intelligent extensions to our brains.

4. TWO DISTINCT MODELING CULTURES AND MACHINE INTELLIGENCE

A famous paper by Leo Breiman characterized the predictive analytics profession as being composed of two different cultures that differ in terms of whether prediction versus explanation is their ultimate goal.[38] A recent paper by Galit Shmueli reviews this same explanation versus prediction distinction.[39] On the one hand, we have purely predictive models like stacked ensemble models that do not allow for causal explanations. Even in cases where the features and parameters are somewhat transparent, pure prediction models have too many arbitrarily chosen parameters for causal understanding. Pure predictive models are most successful in today's machine learning and artificial intelligence applications like natural language processing as in the example of Watson where the predicted probabilities were very accurate and not likely to be arbitrary. Pure prediction models are suspect when the predicted probabilities or outcomes are inaccurate and likely to change across modelers and data samples. Even the most accurate pure predictive models only work when the predictive environment is stable, which is seldom the case with human social behavior and other similarly chancy real-world outcomes for very long. Sometimes we can update our predictions fast enough to avoid model failure due to a changing environment. But more often that is simply not possible because the model's predictions are for the longer term, as we cannot take back a 30-year mortgage loan once we have made the decision to grant it. Thus, there will be considerable risk when we need to rely upon an assumption that the predictive environment is stable. This point was the thesis in Emanuel Derman's recent book *Models Behaving Badly* which argued that the US credit default crisis of 2008–2009 was greatly impacted by the failure of predictive models with a changing economy.[40] To avoid the risk associated with black box models and environmental instability, there is a strong desire for models with accurate causal explanatory insights.

The problem is that the popular standard methods used to generate parsimonious "explanatory" models like standard stepwise regression methods do not really work to model putative causes because what they generate are often completely arbitrary unstable solutions. This failure is reflected in Breiman's characterization of stepwise regression as the "quiet scandal" of statistics. Like many in today's machine learning community, Breiman ultimately saw no reason to select an arbitrary "explanatory" model, and instead urged focus on pure predictive modeling. Yet, any debate about whether the focus should be in one or the other of these two cultures really misses the idea that the brain seems to have evolved both types of modeling processes as reflected in implicit and explicit learning processes. So, both of these modeling processes should be useful in artificial intelligence attempts to predict and explain, as long as there is a semblance to these two basic brain memory learning mechanisms.

The brain's implicit and explicit learning mechanisms both generate probabilistic predictions, but they otherwise have very different properties relating to the diffuse versus sparse characteristics of the underlying neural circuitry and reliance upon direct feedback in these networks.[41] The brain's implicit memories seem to be built from large numbers of diffuse and redundant feed-forward neural connections as observed in Graybiel's work on procedural motor memory predictions projecting through basal ganglia. As seen in the patient E.P., these implicit memory mechanisms are not affected by brain injuries specific to the medial temporal lobe involving the hippocampus. In contrast, explicit learning seems to involve reciprocal feedback circuitry connecting relatively sparse representations in hippocampus and neocortex structures.[42]

Indeed, explicit memory representations in hippocampus appear to become sparse through learning. For example, a study by Karlsson and Frank[43] examined hippocampus neurons in rats that learned to run a track to get a reward. In the early novel training, neurons were more active across larger fractions of the environment, yet tended to have low firing rates. As the environment became more familiar, neurons tended to be active across smaller proportions of the environment, and there appeared to be segregation into a higher and a lower rate group where the higher firing rate neurons were also more spatially selective. It is as though these explicit memory circuits were actually developing sparse explanatory predictive memory models in the course of learning. But unlike stepwise regression's unstable and arbitrary feature selection, the brain's sparse explicit memory features are stable and meaningful. Humans with at least average cognitive ability can usually agree on the important defining elements in recent shared episodic experiences that occur over more than a very brief period of time[44] or in shared semantic memories such as basic important facts about the world taught in school.

Episodic memory involves the conscious recollection of one or more events or episodes often in a causal chain, as when we remember the parsimonious temporal sequence of steps in a recent experience like if we just changed an automobile tire. Some cognitive neuroscientists now believe that similar explicit neural processes involving the medial temporal lobe are the basis of causal reasoning and planning. The only distinction is that the imagined causal sequence of episodes that would encompass the retrieved explicit memories are now projected into the future like when we plan out the steps that are necessary for changing a tire.[45] In fact, evidence suggests that the hippocampus may be necessary to represent temporal succession of events as required in an understanding of causal sequences.[46,47] In contrast to the brain's explicit learning and reasoning mechanisms, parsimonious variable selection methods like stepwise regression usually fail to select causal features and parameters in historical data due to multicollinearity problems. Because of such multicollinearity error, it is obvious that widely used predictive analytics methods are also not smart enough to reason about future outcomes in simulations with any degree of certainty.

Some may say that it would be impossible to model classic explicit "thought" processes in artificial intelligence because consciousness is necessarily involved in these processes. However, machine learning models that simulate the brain's explicit learning and reasoning processes should at least be useful to generate reasonable causal models. If these machine models provide unbiased most probable explanations that avoid multicollinearity error, then these models would be a realization of Leibniz's *Calculus Ratiocinator*. This is because they would allow us to answer our most difficult questions with data-driven most probable causal explanations and predictions.

5. LOGISTIC REGRESSION AND THE *CALCULUS RATIOCINATOR* PROBLEM

While there are probably hundreds of different predictive analytics methods that have at least some usage today, regression and decision tree methods are reported to be the most popular tools.[48] Of these most popular methods, logistic regression is probably used most widely today, as it has shown very rapid adoption since it began to appear in statistical software packages in the 1980s.[49] Much of the reason for the rapid rise in popularity is that it has effectively replaced older less accurate methods that were based upon linear regression like discriminant analysis[50] and linear probability.[51] Another reason is that it has interpretation advantages in the sense of returning so-lutions that are much more understandable compared with the older probit regression.[52] Yet another reason for this rise in popularity is that logistic regression can be applied to all types of predictive analytics problems including those with binary and categorical outcome variables, those with continuous outcome variables that are categorized into ordinal categories,[53] and those with matched samples in conditional logistic regression models. Logistic regression even has been even proposed for use in survival analysis although it is not yet widely used there.[54] Today, logistic regression is applied across most major predictive analytic application areas in fields like epidemiology, biostatistics, biology, economics, sociology, criminology, political science, psychology, linguistics, finance, marketing, human resources, engineering, and most areas of government policy.

Logistic regression can be derived from the much more general maximum entropy method that arises in information theory; this derivation returns the sigmoid form that describes the probability of a response in lo-gistic regression without having to assume such a form up front.[55] Given this fundamental connection to information theory, the thesis of this book is that this popularity of logistic regression stems from something very basic about how our brains structure information to understand the world. This is that our brains may work to produce explicit and implicit neural learning through a stable process that can be understood through information theory as a logistic regression process. The brain may use logistic regression here because it is the most natural calculus to compute accurate probabilities given binary neural signals. That is, there simply are not many choices for how probabilities of binary signals might be computed, and if we limit ourselves to information theory considerations then we arrive at logistic regression.

So in spite of all the limitations in not being able to handle high-dimension and multicollinearity problems, it still might be that we have discovered the essence of a basic probabilistic learning process that represents information at the most basic binary neural level. The heart of this argument is that any cognitive machine designed to predict and explain our world will make the most sense if it employs the same most basic cognitive design principle as our brains. Of course, to buy into this argument one has to accept that our brains produce cognition based upon neural computation mechanisms that are essentially logistic regression, and one major purpose of this book is to lay out this case. Before making this case though, it is worthwhile to point out that it would be most natural for us to understand our world through artificial intelligence devices that are neuromorphic or designed to simulate the very binary information representation process that happens in real neural ensembles in the brain. And that is what we may have been doing in applying logistic regression to a wide variety of practical and scientific problems since the 1980s. This is in spite of shortcomings due to an inability to deal with high-dimension multicollinear data.

This book will propose a slight modification to standard logistic regression called Reduced Error Logistic Regression (RELR) that does overcome these problems. RELR changes logistic regression so that those error properties most famously elucidated by Nobel Laureate Daniel McFadden are now explicitly modeled as a component of the regression. Because evidence reviewed in this book suggests that neurons are especially good at learning stable binary signal probabilities with small sample data given high-dimension and/or multicollinear features, RELR has the effect of making logistic regression a much more general and stable method that is proposed to be a closer match to the basic neural computation process. Note that RELR is pronounced as "yeller" with an r or "RELLER".

The theory of neural computation and cognitive machines that is the basis of RELR is based upon the most stable and agreed upon aspects of information theory. RELR results from adding stability considerations to information theory and associated maximum likelihood theory to get a closer match to what may occur in neural computations. This theory is concerned with stability in all aspects of causal and predictive analytics. This includes stability in regression estimates and feature selection in pure prediction models, stability in regression estimates and feature selection in putative explanatory models, stability in updating the regression estimates in online incremental learning, and stability in regression estimates in causal effect estimates based upon matched sample observation data.

In the end, this book will propose that RELR is a viable solution to the Leibniz's long-standing *Calculus Ratiocinator* problem in the sense that it can simulate stable neural computation, but produces automatic maximum probability solutions that are not corrupted by human bias. RELR can be used in all the same applications as standard logistic regression is used today. Yet, because RELR avoids the multicollinearity problems, this book will argue that RELR will have much wider application because its solutions are much more stable. The slight modification to standard logistic regression that is RELR simply ensures that error in measurements and inferences are effectively taken into account as a part of the basic logistic regression formulation. In the terms of regression modeling theory, RELR is an error-in-variables type method[56] that estimates the error in the independent variables in terms of well-known properties instead of assuming that such error is zero.

Modifications to standard logistic regression in the newer penalized forms of logistic regression also attempt to correct these multicollinearity deficiencies, although these are not error-in-variables methods because unlike RELR they do not estimate the error in the independent variables. Instead, LASSO and Ridge logistic regression and other similar methods employ penalty functions that smooth or "regularize" error and have the effect of shrinking regression coefficients to smaller magnitudes than would be observed in standard logistic regression.[57] These shrinkage methods may require that a modeler uses cross-validation through an independent validation sample to determine the degree of shrinkage that gives the best model, or they might even employ a predetermined definition of "best model" without any cross-validation. In any case, the "best model" will be an arbitrary function unless that best model is the maximum probability model which none of these penalized methods are designed to estimate. Yet, the most fundamental problem with the penalized methods is that there is no agreement on what the penalizing function should be as LASSO has it as a linear function, whereas Ridge has it as a quadratic function, whereas Elastic Net combines both linear and quadratic functions, and there are an endless number of possibilities. These different penalizing functions usually give radically different looking models.[58]

Unlike the arbitrary penalized forms of logistic regression, RELR does not require any arbitrary or subjective choices on the part of the model builder in terms of how to deal with error parameters or error functional forms or "best model" definitions. Rather than allow the effect of error to be happenstance as in standard logistic regression where it will have a very large debilitating effect on the quality of the model's ability to predict and explain with high-dimension or multicollinear data, RELR effectively models this error and largely removes it. The end result is that RELR gives reliable and valid predictions and explanations in exactly those error-prone situations with small observation samples and/or multicollinear high-dimension data where standard logistic regression fails miserably and may not even converge.

RELR can generate sparse models with relatively few features or more complex models with many redundant features. *Implicit RELR* is a diffuse method that usually returns many redundant features in high entropy models.[59] *Explicit RELR* is a parsimonious feature selection method that returns much lower entropy models due to conditions that optimize sparse features.[60] Both Implicit and Explicit RELR methods are maximum likelihood methods, but their maximum likelihood objective function differs depending on whether diffuse versus sparse feature selection is the goal. Much of this book starting with the next chapter explores the statistical basis of Implicit and Explicit RELR in maximum likelihood and maximum entropy estimation theory, along with the unique applications of each of these RELR methods and their putative analogues in neural computation.

RELR avoids the "no free lunch" problem because RELR changes the goal of machine learning. Instead of optimizing an arbitrary classification accuracy or cost function like in all other predictive analytic methods, the objective in both Explicit and Implicit RELR is defined through maximum likelihood criteria when both outcome and error event probabilities are included in the model. It is recognized that because RELR solutions are most probable solutions under the maximum likelihood assumptions, they will not always be the absolute most accurate in every situation. That is, there is always the possibility that less probable models will be more accurate in any given situation. Yet, available evidence is that RELR models are accurate and reliable models that generalize well and which can be automatically generated without human bias. This is all that can be demanded of any machine learning solution.

Although Explicit RELR's feature selection learning may be produced very rapidly in today's massively parallel computing, it is still not a process that ever could be realized in real time. This is because it requires feedback to guide its explanatory feature selection, and this is the small percentage of its processing that cannot be implemented in parallel. On the other hand, Implicit RELR can generate rapid feed-forward learning that is not quite real-time but still very fast in embarrassingly parallel Big Data implementations that require no feedback. In this sense of the relative speed of processing and the extent to which feedback determines each step in the process, this book will propose that Explicit RELR is similar to slow, deep, explicit memory learning, whereas Implicit RELR is similar to fast, shallow, implicit memory learning in neural computation.

So we will learn that the causal explanatory feature selection process guided by Explicit RELR is clearly a slower process than the rapid purely predictive Implicit RELR, and this is also the hallmark of explanatory reasoning compared with predictive intuition. Although Explicit RELR's most probable explanatory models are automatically produced in the computational trajectory that leads to an optimal solution, we will also see that their putative causal features require independent experimental testing. This is because these causal explanations produced are at best putative and most probable causal descriptions that clearly can be falsified with further and better data.

Explicit RELR's hypothesized explanations can be tested with randomized controlled experimental methods or with quasi-experimental methods that attempt to match observation samples to adequately test the validity of putative causal factors. However, quasi-experimental propensity score matching methods also suffer from many of the same estimation problems that have plagued variable selection in predictive analytics, as they also typically employ the same problematic unstable standard stepwise logistic regression or decision trees. In this regard, we will see that RELR gives rise to a new matched sampling quasi-experimental method that is not a propensity score matching method and instead is an outcome score matching method. RELR's outcome score matching quasi-experimental method holds all other factors to their expected outcome probability while varying the putative causal factor. Since the human brain does not perform randomized controlled experiments but obviously can discover causes, this book argues that a similar matched sample method could be the basis of the brain's causal reasoning. This quasi-experimental outcome score matching is an important aspect of RELR's *Calculus Ratiocinator* cognitive machine methodology.

One view of the origin of a science is that it begins with technical innovation in a craft that has large practical application. This view holds that

it is the attempt to make theoretical sense out of such technological innovation that generates scientific principles rather than vice versa. This logical
positivist view was most famously held by the Austrian physicist Ernst Mach
who doubted the existence of atoms[61] and the twentieth century behavioral
psychologist B.F. Skinner who believed that psychology should develop
without an attempt to understand cognition.[62] Mach argued in the *Science of
Mechanics* that the understanding of the concept of force came from our
understanding of the effect of levers, so it is the practical invention that
guides theoretical science rather than vice versa.[63] People who work in
predictive analytics, cognitive science and artificial intelligence tend to be
much more the craftsperson than the theoretical scientist, so there is almost
no underlying theory in this evolving field. Indeed, standard logistic
regression with all of its limitation has had significant practical application in
the past 30 years, but there has been no fundamental theory that explains
why it fails so miserably with multicollinearity and how to correct this
limitation. In fact, RELR grew out of my years of business and science
analytics practice with no real guiding theory except to try to understand
this multicollinearity limitation in standard logistic regression and other
predictive analytic methods.

   Without underlying theory, predictive analytics is in danger of being
viewed somewhat like alchemy with many arbitrary techniques that
sometimes produce an interesting result, but which also sometimes have
significant risk of danger. Just as medieval alchemists in China accidentally
discovered gun powder[64] when they sought an elixir that would allow
potency and immortality, predictive analytics today also can be accidental
and random and produce risky results that are opposite to our intentions as
in the 2008 United States financial crisis. Just as with alchemy, our problem
is that we have no theory to guide our quests. Alchemy's methods to turn
base metals into gold were ultimately shown to be inconsistent with atomic
theory. Similarly, Mach's extreme avoidance of atomic theory was proven to
be impractical, as one cannot imagine the development of chemistry,
physics, chemical engineering, biochemistry, electrical engineering and
even neuroscience without the concept of atoms.

   In today's predictive modeling applications, it is very difficult to describe
sophisticated approaches to prediction and explanation without using the
neuromorphic term "learning". Whether it is called statistical learning or
machine learning, we are modeling a high-dimension multicollinear
computation process that likely would be similar to what occurs in neurons
if it is stable and accurate. So, the neuron is the fundamental structure in

much the same way that the atom is the fundamental structure in physical science. Thus, the theory of RELR that emerges as a basis for *Calculus Ratiocinator* cognitive machines is a view based upon neuroscience. The crux of this theory is that the same maximum entropy concept that Boltzmann derived to characterize the maximum probability behavior of molecules also characterizes logistic regression neural computation.

CHAPTER 2

Most Likely Inference

"A given system can never of its own accord go over into another equally probable state but into a more probable one..."
Ludwig Eduard Boltzmann, "The Second Law of Thermodynamics", 1886.[1]

Contents

Boltzmann's concept of the Second Law of Thermodynamics necessarily assumed the existence of atoms. As a result, he spent much of his career arguing with the majority of his peers like fellow Austrian physicist Ernst Mach who vehemently disputed the existence of atoms. Boltzmann committed suicide in 1906 at the low point of atomic theory acceptance. Ironically, in just a few years after Boltzmann's death, experimental evidence was finally produced that led to the undeniable conclusion that atoms existed.[2] Yet, it took many years after that for the atom to become part of the paradigm theory in normal science education. My father went to high school in a small town in the Midwest of the United States in the early 1940s where his elderly science teacher still taught that the existence of atoms was a controversial idea.

The Second Law of Thermodynamics is one of the great achievements of modern science. The modern understanding of this law is somewhat different from Boltzmann's understanding quoted above. It is that ergodic[3] physical systems when isolated should be determined by laws of probability and be most likely to exist in maximum probability states. So, in the current understanding, a system may of its own accord go into other states besides a maximum probability state, but the maximum probability state is substantially more likely in complex physical systems with many particles such as ideal gases. The central concept of the second law is entropy as the most

27

probable state is maximum entropy. The Boltzmann entropy expression is as follows:

$$S = k \log W \tag{2.1}$$

where S is the entropy, k is a scaling constant, and W describes the number of ways that particles can be arranged in terms of positions and momenta in space and time according to basic laws of combination. A system with greater entropy has more possible arrangements, so with greater entropy there is less information or more uncertainty in the actual arrangement.

In the middle of the twentieth century, Shannon was able to show that uncertainty is generally measured by an equation that has the same form as Boltzmann's entropy.[4] Using the base e logarithm, the discrete form of Shannon's entropy H for a probability distribution $\mathbf{p}$ takes the form:

$$H(\mathbf{p}) = -\sum_{j=1}^{C} p(j)\ln(p(j)) \tag{2.2}$$

where $p(j)$ is the probability of an event across C possible outcome categories. So, in the case of a coin flip, C would be equal to 2 as these are just two possible outcomes. In that case, it is easy to verify that $H(\mathbf{p})$ would be maximal when $p(j)$ is equal to 0.5 for each of these two possible outcomes where the sum is constrained to equal 1. In that fair coin case where each of the two possible outcomes has equal probability, $H(\mathbf{p}) = 0.693$. On the other hand, $H(\mathbf{p}) = 0.325$ when one $p(\text{heads}) = 0.1$ and $p(\text{tails}) = 0.9$ or vice versa in biased coins. In general, $H(\mathbf{p})$ is always less than 0.693 when $p(\text{heads})$ differs from $p(\text{tails})$. So, there is a maximal value of uncertainty $H(\mathbf{p})$ about the outcome in a single coin flip when a head and a tail are equal in probability which makes sense. Shannon realized that with less uncertainty, there is always more information. Shannon's entropy expression is one of the cornerstones of information theory.

1. THE JAYNES MAXIMUM ENTROPY PRINCIPLE

Soon after Shannon, Edmund Jaynes came to believe that because maximum entropy also reflects maximum probability, this suggests a very fundamental principle. In essence, he reversed the statistical mechanics problem that assumed that maximum entropy would be the most reasonable solution under assumptions about ergodic behavior in isolated systems and simply asked "what is the most probable behavior of a system given our historical observations and known constraints?" The answer Jaynes gave was:

> ... we take entropy as our starting concept, and the fact that a probability distribution maximizes the entropy subject to certain constraints becomes the essential fact which justifies the use of this distribution for inference.[5]

But Jaynes argued that this maximum entropy inference has much more general application than just in statistical mechanics, as he believed that the same principle always should be applied in every well-reasoned inference. In this general sense, Jaynes proposed that maximum entropy inference is always as much about the uncertainty that we have about our world than about anything else. Even to the extent that we can be certain in our constraints, we also must accept that there are other uncertainties related to what we do not know. So our measurements are likely to be incomplete because they are based upon finite data or do not include all constraints and thus there will be uncertainty in our knowledge. Yet, by maximizing entropy subject to certain constraints, this allows a most probable solution given incomplete information. The fundamental assumption is that all observations must be independent. This general principle of inference is now called the *Jaynes maximum entropy principle.*[6]

An understanding of this principle can be found in the simple example of inference that is possible in the case of the behavior of water molecules. Say that you have a closed container with liquid water at equilibrium where we measure the temperature. Now say that we open the system and add heat until the temperature reaches boiling, and the system is then closed again and at equilibrium. The added heat has caused molecules on average to increase in velocity, but we cannot say that any individual molecule has increased in velocity. This is due to the variability at any one point in time because some molecules might even be stationary as they stop momentarily when they run into each other. But at least in this case, one can characterize the most probable behavior of molecules by seeking the maximum entropy solution given the heat energy. We can develop probabilistic models that predict on average how an independent causal variable like heat may impact the probability that a given molecule may have specific behavior. This characterization only will be probabilistic and apply to average group behavior and not to individual molecules with certainty. But this can be even a causal inference under well-controlled comparison conditions that hold everything constant except for one changing causal factor. The tremendous insight of Jaynes was that this same principle could apply very generally to give most likely inference in all situations where uncertainty exists, and this even would include open systems and human behavior if applied correctly.

A naïve understanding of entropy assumes that a maximum entropy solution subject to all known constraints is necessarily a highly disordered pattern. But snowflakes are examples of a maximum entropy distribution of molecules subject to constraints. In snowflakes, a static maximum entropy distribution is frozen on the way to thermodynamic equilibrium that obeys constraints related to interactions between molecules and rate of freezing.[7] In this case, these constraints impose strikingly low-entropy or high-information structure upon the resulting maximum entropy configuration. This low-entropy pattern is seen in the simple whole number harmonic ratios such as 2:1, 3:1, and 4:3 in the relative counts of the spikes and/or sides of the various embedded figures (Fig. 2.1).

Snowflakes are formed in open systems where entropy in molecular configurations is forced to decrease due to local environmental constraints. Biological systems like neural systems are also open systems, although like

Figure 2.1 One of the snowflakes photographed by Wilson Bentley. Notice the simple whole number harmonic ratios such as 2:1, 3:1, and 4:3 in the relative counts of the spikes and/or sides of the various embedded figures.[8]

snowflakes they also can be assumed to exhibit stable maximum entropy phases subject to known constraints. For example, the Boltzmann maximum entropy distribution equation describes how the concentration gradient in the distribution of a charged molecule across the neural membrane is related to an electrostatic potential gradient in Hodgkin and Huxley's original explanation of the resting potential behavior of the neuron.[9] Like the snowflake, this effect can be understood as a mechanism that maximizes entropy subject to known constraints. Like the snowflake, this is a clear example of the Jaynes principle that gives a most probable inference that explains a physical phenomenon.

In the neuron, most probable entropy decreases well beyond what would be expected in random behavior also can be represented as due to fundamental constraints in a maximum entropy inference system. For example, the membrane potential changes at the axon hillock that are graded potentials and change the probability of axonal spiking are caused by highly targeted synaptic and dendritic effects. Later in this book, it will be seen that these effects can be modeled as constraints in a maximum entropy formulation that is the RELR neural computation model. When these constraints are stable in the sense that the input features are fixed, the system will be most likely to produce the maximum entropy state subject to these input constraints. Later in this book, it also will be seen that very high information, small whole number frequency ratios of oscillating neural ensemble patterns can arise temporarily in the human brain during explicit cognition. These high information patterns are analogous to the low-entropy patterns of snowflakes.

In physical systems such as snowflakes and vibrating musical strings, simple harmonic relationships are easily understood as arising from stable orthogonal components in whole number frequency ratio relationships and easily analyzed with simple tools that assume that the harmonic frequencies fit perfectly within very regular analysis windows. But in complex systems that produce cognition and behavior, the problem is that perfect harmonic relationships are much more transient if they even exist. So, simple tools that force an assumption of stable low-entropy orthogonal relationships that fit perfectly within an analysis window like principal components analysis or Fourier analysis are not really effective and are more likely to force such relationships to exist when they are not in fact there. It is much better to use a method that does not force low-entropy structures that do not exist but will discover such structures if they do in fact exist, and the maximum entropy method is such a method.

2. MAXIMUM ENTROPY AND STANDARD MAXIMUM LIKELIHOOD LOGISTIC REGRESSION

Although standard maximum likelihood logistic regression is probably the most popular approach in today's predictive analytics practice, many analysts do not know that it gives identical solutions to the maximum entropy subject to constraints method.[10] Yet, the predictive features in standard logistic regression may be unlikely to be certain constraints unless determined through experimental means such as discrete choice methods. Thus, it probably would be a stretch to say that the widespread practice of standard logistic regression is always an application of the Jaynes principle of most likely inference.

In maximum likelihood logistic regression, observed outcomes in the form of binary dependent variable events are used to compute a function that estimates their probability of occurrence given the predictive features. This probability function is the joint probability that all observed outcome events occur. This is simply the product of the estimated probability of all observed outcomes just like the joint probability of independent coin flip trials is a product of the probability of all those individual outcome events. For example, the joint probability of two straight heads in a fair coin flip is $0.5 \times 0.5 = 0.25$, whereas the joint probability of three straight heads is $0.5 \times 0.5 \times 0.5 = 0.125$. Again, this simple multiplicative joint probability rule does assume that the outcome events are independent events, so one event cannot cause or depend upon another event.

In software implementations, the logarithm of the likelihood is maximized in logistic regression because it is easier to compute. This Log Likelihood or LL of a probability distribution $\mathbf{p}$ is written as:

$$LL(\mathbf{p}) = \sum_{i=1}^{N} \sum_{j=1}^{C} y(i,j) \ln(p(i,j)) \qquad (2.3)$$

where $y(i,j)$ is the outcome for the ith observation in the jth category condition which is coded as "1" when an event occurs and "0" when an event does not occur. Similarly, $p(i,j)$ is the estimated probability for the ith observation in the jth category condition. In a maximum likelihood solution, each $p(i,j)$ is computed so that $LL(\mathbf{p})$ is maximized. Notice that the log likelihood is simply the logarithm of the product or joint probability of all observed outcome events which are assumed to be independent.

For the binary case, there are two category conditions, so $j = 2$. So, with two coin flips, an instance of an event such as getting a head on a trial can be coded as "1" and a noninstance of an event such as not getting a head can be coded as "0". If a head is observed on the first coin flip trial and a tail on the second, then the sequence is coded in terms of $y(i,j)$ as 1, 0, or head/nontail for the first event and 0, 1 or nonhead/tail for the second event. In this case, then $LL(\mathbf{p}) = 1 \times \ln(0.5) + 0 \times \ln(0.5) + 0 \times \ln(0.5) + 1 \times \ln(0.5) = -1.386$ because the maximum likelihood estimation would yield equal probability estimates $p(i,j)$ for heads and tails. Unlike Shannon's entropy expression in Eqn (3.1) which always gives positive values, this log likelihood expression in Eqn (3.2) always gives negative values. Yet, just like the entropy measure, any probability estimates whereby p(heads) $\neq p$(tails) would give lower log likelihood values. Because the probabilities that the maximum likelihood estimation generates are driven by empirical outcome event observations, they can be inaccurate in small or unrepresentative samples.

For a direct comparison to this log likelihood coin flipping example, in a binary outcome across N trials, Shannon's entropy can be written as:

$$H(\mathbf{p}) = -\sum_{i=1}^{N} \sum_{j=1}^{C} p(i,j)\ln(p(i,j)) \tag{2.4}$$

where $p(i,j)$ is the outcome probability for the ith observation in the jth category condition. In a maximum entropy solution, each $p(i,j)$ is computed so $H(\mathbf{p})$ is maximized. If we take the case with two coin flips giving one head and one tail, we can compute the negative sum that gives the entropy as $-0.5 \times \ln(0.5) + -0.5 \times \ln(0.5) + -0.5 \times \ln(0.5) + -0.5 \times \ln(0.5) = 1.386$. So, notice that the entropy and log likelihood have the same magnitude, but opposite signs for the same probability distribution. This will be generally the case.

As in this example, it does not matter whether the entropy of the estimated probability distribution is maximized subject to constraints or the log likelihood is instead maximized. The very same estimated probability distribution simultaneously maximizes both objectives. With the exception of certain pathological cases, both maximum entropy and maximum likelihood logistic regression are also well-behaved optimization objectives where any local optimum is also a global optimum. One pathological condition is where perfect discrimination accuracy with zero error can be observed known as complete or quasi-complete separation, depending upon whether ties exist.

In this case, there is no longer any unique optimal solution.[11] Another pathological condition is that standard maximum likelihood logistic regression often fails to converge in the face of significant multicollinearity. We will see that RELR largely overcomes both of these pathological cases.

Whereas Shannon's entropy can be interpreted to measure uncertainty, a reasonable question is what does log likelihood measure in logistic regression? Since it is opposite in sign to uncertainty, log likelihood might be thought to be a goodness of fit measure which reflects the extent which a probability distribution matches the observed outcomes. Log likelihood gets greater or less negative as this match gets better and better. In general, a probability distribution which has a relatively higher or less negative log likelihood value would be a better match to data than one with a relatively lower or more negative log likelihood value. On the other hand, because entropy is opposite in sign to log likelihood, a higher value of entropy would be a poorer match to data and would be associated with greater uncertainty. So even though maximum entropy and maximum log likelihood generate the same probability distribution for the very same fixed set of predictive features, it is important to recognize that entropy and log likelihood have different meanings and will give different optimal values across changing predictive feature sets. For example, feature selection methods can be designed to discover the optimal objective function across many possible predictive feature sets, and the maximum entropy and maximum log likelihood solutions across these many possible feature sets normally would be different. In such feature selection optimization, both Implicit and Explicit RELR optimize maximum log likelihood functions for obvious reasons that are reviewed in the next chapter.

No matter whether it is computed through maximum log likelihood or maximum entropy, in logistic regression, the function that determines the estimated probability for each particular binary target outcome event follows a logistic distribution. That is, the probability of a binary outcome event at the ith observation and $j = 1$ and $j = 2$ categories is determined as follows:

$$p(i, j = 1) = \frac{e^{\alpha + \sum_{r=1}^{M} \beta(r)x(i,r)}}{1 + e^{\alpha + \sum_{r=1}^{M} \beta(r)x(i,r)}} \quad \text{for } i = 1 \text{ to } N, \quad (2.5a)$$

$$p(i, j = 2) = \frac{1}{1 + e^{\alpha + \sum_{r=1}^{M} \beta(r)x(i,r)}} \quad \text{for } i = 1 \text{ to } N, \quad (2.5b)$$

where $x(i,r)$ is the value of the rth feature at the ith observation, $\beta(r)$ is the regression weight for the rth feature, and α is an intercept.[12] The assumption that the distribution function is logistic is not the only way to get a maximum likelihood regression model with binary target outcomes. However, logistic regression is generally considered to be as accurate or more accurate and easier to compute than these other possibilities. Another older approach that is still sometimes used in econometrics is probit regression which assumes that this distribution function is normally distributed. Interestingly, probit regression and logistic regression give probability estimates that are almost indistinguishable,[13] so standard logistic regression is usually used in this case simply because it is far easier to compute and much easier to interpret. Yet, logistic regression also has a basis in information theory through its equivalence to maximum entropy, and this might be its greatest theoretical advantage.[14]

The practical interpretation advantage of logistic regression is seen in the nice linear form of the logit or log odds, as probit regression has no comparable way to interpret odds so easily. From the probability value $p(i,j)$, this logit or log odds of a binary event in the ith observation and jth category easily can be derived which is

$$\text{logit}(i,j) = \ln\left(\frac{p(i,j)}{1 - p(i,j)}\right) \quad \text{for } j = 1 \text{ to } 2 \text{ and } i = 1 \text{ to } N. \quad (2.6)$$

or for $j = 1$,

$$\text{logit}(i,1) = \alpha + \sum_{r=1}^{M} \beta(r)x(i,r) \quad \text{for } i = 1 \text{ to } N. \quad (2.7)$$

This logit definition in Eqn (2.7) has the same form as an ordinary linear regression equation although there is no error term. Like linear regression, the features in the logit are independent variables that combine in a linear additive manner and also can be nonlinear and interaction effect terms, so a logit model that arises from logistic regression is not limited just in terms of linear main effect features. Like ordinary linear regression, a problem with standard logistic regression is that overfitting error and related multicollinearity error occur as more features are added, and this can be especially problematic with nonlinear and interaction effects. When overfitting and other related multicollinearity error occur in standard logistic regression applications, such error is a sign that the Jaynes principle that maximum entropy gives the most probable inference cannot be applied. This is because the Jaynes principle only applies to the case where certain constraints are entered into a model and overfitting error is clearly a case where unknown or unfounded independent variable constraints are entered in a model. In reality, there is almost always significant error in the logit estimate in standard logistic regression due to incorrect data values, multicollinearity, overfitting, omitted variables, missing data, sampling, outliers, and rounding. While this logit error is not a parameter that is ever estimated in standard logistic regression, it still has been intensively studied in one major application area. This is discrete choice modeling.

3. DISCRETE CHOICE, LOGIT ERROR, AND CORRELATED OBSERVATIONS

Discrete choices are decisions involving two or more alternatives such as whether to vote for the Republican or Democratic candidate, whether to take a plane or ride a bus, or whether to take a job or not. While not all person attributes can be experimentally manipulated in a choice study, a researcher often can vary the attributes associated with a choice experimentally such as the price or quality of product to be purchased, and determine causal effects on the decisions. At least with these choice attributes where randomized experiments are possible, reasonably certain predictive features may be selected in a model in accord with the Jaynes principle. However, besides predictive features that a researcher can test and select experimentally, it is likely that other important unobserved constraints will have an impact. So, discrete choice models are subject to significant logit error problems due to unobserved constraints.

Discrete choices are most often assumed to be based upon a process whereby decision makers choose an alternative that maximizes their utility or net benefit given the attributes of a decision. One view known as causal decision theory is that such utility directly reflects causal determinants of behavior, whereas another evidential view is that it only reflects a conditional probability which is sometimes only correlational. Given the very same decision variables and historical learning, these two utility processes often generate the same optimal decisions. However, there are some interesting scenarios where the evidential theory would predict that you would do that which you would be happiest to learn that you have done, whereas the causal theory would predict that you would do that which is most likely to bring about good results.[15] Causal decision theory is more consistent with decisions that are under conscious or explicit control.

Evidential decision theory is more consistent with decisions that are not under conscious causal control. Still, unconscious or implicit choices might be immediately available to consciousness after they are made. The view of this book is that both explicit and implicit processes are important in all phases of human cognition including decisions, so the present view is that both causal and evidential decision theories may each reflect an aspect of decision processing. However, decisions that are guided by causal understanding would be viewed as better decisions and less risky, although there are certainly exceptions related to the need for speedy decisions in stable environments where causal understanding is not such an important factor.

Under either the causal or the evidential theory, if a person is given a choice between taking a plane or riding a bus to a distant city, and if the person is poor and only has the money for a bus ride and also has the time to ride the bus because they are unemployed, the utility associated with the bus alternative would be likely to be higher than that associated with the plane. So, if a researcher observed such a person's decision and also was certain of their employment status and savings, then one could predict that the bus alternative would have greater utility than the plane alternative. The problem arises when factors that determine a decision are not observed, as there may be error in the model's prediction due to unobserved factors.

Since the pioneering work of Nobel Laureate Daniel McFadden,[16] discrete choice has been traditionally modeled through standard logistic regression where the alternatives are response categories and the choice and decision maker attributes are predictive features. The utility or $U(i,j)$ that the ith decision maker assigns takes the identical form of the logit[17] so for the $j = 1$ binary category alternative in a discrete choice model, this becomes

$$U(i, 1) = \varepsilon(i, 1) + \alpha + \sum_{r=1}^{M} \beta(r)x(i, r) \quad \text{for } i = 1 \text{ to } N. \quad (2.8)$$

This is exactly the same form as Eqn (2.7) and the indices have similar meaning, but there is the addition of the error term $\varepsilon(i,1)$. In the most general form that still maintains the linear additive characteristics of each individual feature value $x(i,r)$ being multiplied by a regression weight $\beta(r)$ and then summed, the features in the choice model given by the vector x may be attributes of a choice, attributes of a person, along with interactions and nonlinear effects. In the binary discrete logit choice model with just two categories, the error term $\varepsilon(i,1)$ is not estimated directly and is treated as random just like how ordinary linear regression does not estimate its error term and treats it as random. It represents the errors associated with the jth $(j = 1)$ binary alternative relative to the reference alternative $(j = 2)$ for the ith decision maker. As suggested above, standard logistic regression discrete choice modeling assumes that such modeling error would arise due to the failure to include certain constraints, or what may be called unobserved factors that influence an outcome, along with other forms of measurement error. There is almost always uncertainty because the discrete choice probability estimates seldom match the actual choice responses closely for all observations, so the error term $\varepsilon(i,1)$ is designed to represent this uncertainty formally. Figure 2.2 shows a standard logistic regression model's probability estimates $p(i,j)$ for the $j = 1$ binary alternative across $i = 1$ to N observations, along with the actual binary choices 0 or 1. There is a nontrivial amount of error in the logistic regression model's probability estimates for these choices, as the regression model only provides a best or maximum likelihood fit.

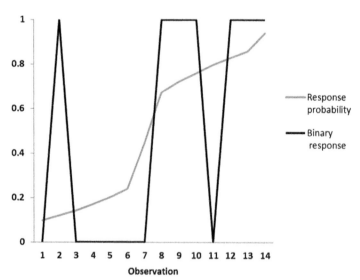

Figure 2.2 Example of substantial error that is usually seen in logistic regression models. The gray curve reflects the probability of the responses, whereas the black curve reflects binary (0, 1) responses across 14 consecutive trials in simulated data. The error that is defined in discrete choice modeling is represented by the difference between these two curves.

A critical assumption in discrete choice standard binary logistic regression modeling is that the modeling error that results from unobserved factors in the $j = 1$ alternative is independent from that in the $j = 2$ alternative.[18] This assumption that the error is independent in the $j = 1$ and $j = 2$ binary choice alternatives would seem realistic in cases where only one observation is measured from each decision maker in a sample of independent decision makers who have no influence upon one another. In such a case, unobserved factors would be present and independent across the two observed binary outcomes because all observations would be independent.

This assumption of independent error might be problematic in other situations where observations are not entirely independent such as when multiple responses are sampled from the same individual decision makers over time. For example, it might be that later responses tend to be only in one alternative versus another due to a drifting response bias where error in one alternative is not independent of error in another alternative. So in standard logistic regression binary choice modeling, the assumption of independent error can be restrictive in situations where there are unobserved causal factors that produce significant serial correlation in response data as reviewed by Kenneth Train.[19]

Unfortunately, the attempts to control for correlations in observations without any sense of causal relations can lead to incorrectly specified and problematic models. For example, the Generalized Estimating Equation (GEE) method has been used as a way to handle correlated observations in logistic regression in what are termed population average models in the biostatistics literature.[20] Yet, this method requires that a researcher specifies the correlation structure between observations, or it will automatically generate such correlation structure with a resulting unwieldy number of parameters with many likely spurious correlations. Because of such large number of parameters, many researchers prefer to specify this correlation with very simple structures that are constant over time. It has been argued that incorrectly specified correlations in GEE will hurt the efficiency of the model, yet it is also argued that the model will still be consistent in the sense that with enough data, the solution will be the same as that which would be obtained with correctly specified correlation structure. Still, there is no guidance on how much data is needed for such consistency. In fact, there can be suspicious problems in GEE in epidemiological studies which are likely due to incorrectly specified time varying covariates.[21]

Other popular methods to deal with error due to correlated observations in standard logistic regression involve some aspect of fixed effect or random effect designs. In these cases, assumptions must be made about whether to treat error due to correlated observations across repeated measures as fixed versus random with strict assumptions involved in each case. When fixed effects are assumed, intercepts or equivalent differencing must be introduced to account for correlated errors due to the distinct clustering units such as stores, schools, and hospitals. These estimated effects along with their interactions can have substantial error. Even more problematic is that fixed effect designs do not allow the estimation of between-subject explanatory effects. For example, between-subject effects like gender effects cannot be assessed as fixed effects in a longitudinal study of observations from the same persons over time because fixed effects are only sensitive to within-person effects that change over time in such a design. In fact, only within-subject effects that change significantly over time can be estimated. But, within-subject effects typically do not have the same variability as between-subject effects, and this makes their estimation more difficult.

As is commonly used, a random effects design usually means that a normally distributed random intercept with a mean of 0 and standard deviation of 1 is forced to occur across higher level clustering units such as schools, although any distribution is theoretically possible.[22] When random effects are assumed in logistic regression modeling, lower level between-person explanatory effects like related to characteristics of students can be estimated but only with the very restrictive assumption that these effects are uncorrelated with the random effects used to model higher level clustering units such as schools. Otherwise, bias will be present. These biased estimates are also not consistent, so they would not be corrected with larger samples that approach the population. This random effects assumption can be quite controversial. Yet, mixed effects designs which require researchers to assume aspects of both fixed and random effects are frequently used with observation data even when the random effects assumption is dubious. This is probably because there have been no other good options.[23]

Kenneth Train comments that the assumption in discrete choice models of independent errors over time often has severe restrictions.[24] To avoid this assumption, he proposes that one may consider using probit or mixed logit designs in discrete choice models. While probit regression often can lead to reasonable estimates, Train warns that the assumption of normally distributed error can lead to wrong signs in certain types of variables such as price. Train also suggests that the proposed probit computation that potentially allows correlated errors over time is complex, so researchers have more recently simply made the assumption that errors are independent over time in probit models. In contrast, he suggests that the mixed logit does not require special computation that is more complex than other aspects for models that assume that error is not independent. Yet, mixed logit is a random effects model and requires the same assumptions as detailed above and also requires a researcher to specify a probability distribution to characterize the error.[25] Because of this, mixed logit has been described as being best suited for high-quality data[26] which presumably would always have to be experimental data. In addition, there is quite a long list of model specifications in mixed logit necessarily under the control of the researcher that can have a very large effect on model quality,[27] and these could be influenced by human biases.

In many cases, Train argues that unobserved factors that produce serial correlation can be modeled correctly in standard logistic regression with completely independent observations. For example, one binary dependent variable outcome that leads another by one observation or $y(i - 1, j)$ may be introduced into a model and be a putative causal attribute for observed serial correlation such as related to habit formation in the case of positive serial correlation or a variety seeking personality in the case of negative correlation. In the biostatistics literature, these kinds of models are called *transitional models*, which are based upon longitudinal studies where previously measured outcomes can be used as covariates in new cross-sectional or cohort samples where all observations are still independent.[28] These transitional models do not allow the degree of control that is possible with experimentally manipulated factors in randomized controlled experiments. So even in the most controlled situation in discrete choice experiments, some reliance on less controlled observation factors might be necessary if one wishes to avoid unobserved factors. The larger problem is that the error due to unobserved factors is really just one of many possible types of measurement errors that could have a major effect on the quality of the model.

4. RELR AND THE LOGIT ERROR

There is a long tradition in linear regression research on the effect of measurement error in both independent and dependent variables.[29] The error-in-variables problem was also first extended to logistic regression many years ago.[30] This work showed that in realistic scenarios where measurement error would be expected in both dependent and independent variables and unobserved effects are present, both standard linear and logistic regression models would produce biased regression coefficients that do not reflect true regression coefficients even when the constraints that are considered are certain and independent observations are present. This work also showed that such bias would not be alleviated by larger samples. So regression coefficient estimates are not consistent under realistic scenarios where not all independent variable constraints are measured or where measurement error exists in both dependent and independent variables even with perfectly independent observations.

How does this square with the Jaynes principle? It suggests that there are limits to what we can know with standard regression methods even when we know a subset of the true constraints with certainty because our regression estimates will still be biased unless the error is uncorrelated with the measured known independent variable constraints. Like the discrete choice modeling research, these derivations and simulations on error-in-variable effects do not even consider all sources of error. However, this work does include measurement error due to sampling, underfitting error due to unobserved factors, and multicollinearity error in the sense that an unobserved factor can bias regression coefficients in observed factors.[31] Yet, overfitting error is another obvious other source of error, as well as multicollinearity error in measured independent variables. So the news is not good for standard regression modeling in respect to the Jaynes principle because the most likely inference still may not be a very good inference even with perfectly independent observations and certain independent variable constraints in the face of other significant sources of error. The only occasion where standard regression methods might avoid these problems would be where there are orthogonal independent variables and little measurement error in both independent and dependent variables.

In binary logistic regression or ordinal logistic regression with very few ordinal categories, measurement error in the dependent variable is much less of a problem because there are very few dependent variable values. This dependent variable error can be expected to subside approximately as an inverse square root function of the sample size, and is reflected by intercept error.[32] So at least with a large enough sample so that variation in the proportion of target responses in different response categories can be expected to be small, one can ignore intercept error as a significant source of error in logistic regression. Yet with all other sources of error, one needs to assume that the aggregate effect of all error will be uncorrelated with estimated regression effects. Effectively, this assumes that the error will sum to zero in standard logistic regression.

This assumption of no error in independent variable effects is very unrealistic. A much more realistic regression model would estimate parameters which reflect error in the independent variable effects as a component of the regression. Yet, standard error-in-variables methods must make reasonable assumptions about error, or else the models will become an artifact of those assumptions. For example, one problem with common error-in-variables methods in linear regression is that an assumption of the ratio of error in the dependent variable to the error in each independent variable is required.[33] But, this is not usually knowable. So, realistic assumptions that are knowable need to be at the heart of any error-in-variables method like RELR.

As an error-in-variables method, RELR starts with a more realistic assumption than standard logistic regression that significant error always will be present due to a wide variety of unknown error sources, but it specifies error modeling parameters designed to estimate the probability of this error. In fact, standard logistic regression can be considered to be a special case of RELR when all error that RELR models is zero. That is, as reviewed in the RELR formulation presented in Appendix A1, the log likelihood expression for RELR can be written as follows:

$$LL(\mathbf{p}, \mathbf{w}) = \sum_{i=1}^{N} \sum_{j=1}^{C} y(i,j)\ln(p(i,j)) + \sum_{l=1}^{2} \sum_{j=1}^{2} \sum_{r=1}^{M} y(l,j,r)\ln(w(l,j,r))$$

(2.9)

where this log likelihood combines two probability distributions $\mathbf{p}$ and $\mathbf{w}$.[34] The first component is reflected by the left most set of summations in Eqn (2.9) involving the outcome probability distribution $\mathbf{p}$. This is called the "observation log likelihood" or OLL, as it reflects the log likelihood corresponding to the observed dependent variable outcome events $y(i,j)$ across the $i = 1$ to N observations and $j = 1$ to C categories. This first component is identical to the log likelihood expression in standard maximum likelihood logistic regression shown in Eqn (2.3). The second component in RELR is the second set of summations across the error probability distribution **w**. This component is unique to RELR, and is called the "error log likelihood" or *ELL*. It reflects the portion of the RELR model that relates to error modeling. The entire $LL(p,w)$ expression in Eqn (2.9) can be called the *RLL* or the "RELR log likelihood". The observations indexed as $y(l,j,r)$ in the RELR error model are pseudo-observations designed to reflect inferred error across the $l = 1$ to 2 positive and negative error conditions, $j = 1$ to 2 error modeling categories and $r = 1$ to M independent variable feature constraints.[35] As reviewed in Appendix A1, the number of categories C in the left hand summations would only be different from 2 in the case of when RELR models ordinal category dependent variables, but summation across the error modeling categories in the right hand summations always should be for $j = 1$ to 2. This is because RELR's ordinal regression is similar to proportional odds ordinal regression, which has identical properties to binary logistic regression except for multiple intercepts. The proportional odds assumption is required in RELR's ordinal regression.

This RELR error model is in the spirit of the Jaynes principle which assumes that the inclusion of known constraints can greatly aid in reducing uncertainty and thus lead to a more likely inference. The basic idea that motivates RELR is that the constraints that determine the error in the logit are known at least in an aggregate sense. These constraints dictate that one need not be concerned about error in the dependent variable or intercept error which can be assumed to be negligible with a large enough sample. So RELR does not attempt to estimate intercept error, as the error modeling in RELR can be assumed to reflect error related to all sources other than intercept error. Both positive and negative errors are inferred always to be present to some degree. In this formulation, parameters that capture the effect of error are now directly estimated. As derived in Appendix A2, the RELR logit for the $j = 1$ binary outcome at the ith observation can be derived from consideration of the joint probability of this outcome and all errors in variables across all $r = 1$ to M features and can be written as follows:

$$\text{logit}(i, 1) = \sum_{r=1}^{M} \sum_{l=1}^{2} \varepsilon(l, 1, r) + \alpha + \sum_{r=1}^{M} \beta(r)x(i, r) \quad \text{for } i = 1 \text{ to } N,$$

$$(2.10)$$

where it is recognized that the far left hand summations involving the error terms cancel to zero given the definition of the error probabilities provided in the appendix. This is comparable to the standard logistic regression logit defined in Eqn (2.7) except that the logit error is now estimated for the $l = 1$ and two positive and negative errors across the $r = 1$ to M independent variable features, so RELR will almost always return different regression coefficients $\beta(r)$ due to this error-in-variables adjustment. The utility U in Eqn (2.8) is also a logit with an error term similar to RELR's logit in Eqn (2.10), but this error in binary discrete choice utility modeling $\varepsilon(i,1)$ is undefined and not estimated. So RELR decomposes and estimates this logit error for each dependent variable category in terms of positive and negative error components that exist for each independent variable feature, whereas discrete choice has traditionally viewed this error as a quantity that cannot be estimated.

A very key assumption in RELR is identical to that in standard maximum likelihood logistic regression. This is that all observations are independent, so dependent variable responses in one observation cannot cause or depend upon dependent variable responses in another observation. Normally great care must be taken to ensure that this assumption is fulfilled though proper sampling design. RELR's error modeling deals effectively with multicollinearity error in dummy coded variables that represent clusters of correlated observations. Still, in all cases it must be assumed that one outcome observation cannot cause or depend upon another outcome observation. This assumption is not often the case in longitudinal data, but transitional models are possible where covariate effects may be based upon outcomes from previous samples, and an example will be shown in Chapter 4. Because RELR allows very small sample models due to its error modeling, knowledgeable researchers and statisticians can find ways to fit models within RELR's independent observations requirements.

Along with this independent observations assumption, RELR's error modeling directly results from the following three assumptions concerning the logit error that is shown in Eqn (2.10):

1. Across independent variable features, positive and negative logit errors are independent and identically distributed Extreme Value Type I random disturbances with the expected positive and negative errors for each independent variable feature being inversely proportional to t.

2. Across independent variable features, the probability of positive and negative errors is equal.

3. Across independent variable features, the probability of positive and negative errors is not biased by whether features have an odd or even powered exponent in a polynomial expression.

In the first assumption, t is a t-ratio statistic that describes the reliability of the correlation of the rth independent variable feature with the dependent variable. This expected error that is inversely proportional to t for each feature is scaled by the sum of all such t values across all M independent variable features and multiplied by 2 to reflect independent positive and negative error estimates as shown in Appendix A1. This scaling has the effect of normalizing the inverse expected error which becomes the expected proportional effect so that the sum across positive and negative errors for all features is simply unity. The expected logit error serves as a naïve prior error, as the actual estimated error may differ substantially from this expected error. The $1/t$ expected error gives extreme expected values of error when t is small in magnitude, so it is consistent with the Gumbel or Extreme Value type 1 error that has been shown to characterize the logit error in standard maximum likelihood logistic regression discrete choice derivations.[36,37] This logit error arises from the demonstration that the difference between two Gumbel extreme value distributed measures is distributed logistic.[38] In the case of RELR, the positive and negative estimated errors in the logit then can be thought to be the two Gumbel extreme value measures that give rise to a difference which is the actual error which is distributed logistic.

In fact, this $1/t$ expected error fits all known relationships between the Student's t distribution, the logistic distribution, and the Gumbel extreme value distribution to a very good approximation. For example, the choice of t in the RELR error model is not arbitrary, as the Student's t-distribution and the standardized logistic distribution are essentially the same distribution.[39] In addition, as shown in Appendix A1, when the two RELR extreme value distributed variables—positive and negative logit errors for a given feature—have their expected error, their difference gives a standardized RELR logit coefficient for that given feature that is approximately proportional to Student's t.[40] Hence, the choice of expected error as being proportional to $1/t$ across features gives all known relationships between extreme value error variables, the logistic distribution, and the Student's t-distribution. Another choice for the expected error such as $1/t^2$ would not preserve these known relationships and instead give an anomalous result such as a proportional relationship between standardized logit coefficients and t^2 instead of t. This would be inconsistent with the result that the difference between two extreme value variables has a logistic probability distribution. Thus, the choice of $1/t$ in the RELR error model is not arbitrary, although it is only a good approximation given that the Student's t-distribution and the logistic probability distribution are only approximately equivalent.

Pearson correlations are used because they have general application across ratio, interval, ordinal and binary level measurements. For example with ordinal variables, Pearson correlations give identical correlations to Spearman correlations when ranked variables are employed, so the coding of ordinal variables into ranked values is the best practice prior to using such variables. Binary variables conveniently produce identical correlations whether they are interpreted as ordinal variables and coded to reflect a rank with Spearman correlation or ratio variables that reflect a probability of 0 or 1 as with Pearson correlation, so there is no need to code binary variables into ranked variables. RELR standardizes all independent variable features to have a mean of 0 and a standard deviation of 1, and missing values are imputed as 0 after the standardization and after dummy coded missing status variables are also employed. The t value that reflects the reliability of the Pearson correlation needs to be computed prior to imputation. When this is done, the $1/t$ expected error gives greater error for features with more missing values everything else being equal which has validity and is quite different from how mean-based imputation typically works. Note that RELR's handling of missing values does not force an assumption of values that are missing-at-random. Even though the mean-based imputation by itself is based upon this assumption with zero information added so the imputation which guards against incorrect guessing of values, the dummy coded missing status features would be sensitive to structurally missing data that are not missing at random. Thus, these dummy coded missing status features can adjust any associated imputed features accordingly if structural missing data have an effect in a model. In contrast to RELR's imputation, multiple imputation can generate significant error if missing-at-random assumptions are not true.

The t values that reflect the reliability of the Pearson correlation between an independent and a dependent variable are generally suited to the error modeling based upon combinations of ratio, interval, ordinal and binary variables. However, independent sample t-tests that compare mean differences in binary groups give almost identical values with balanced samples.

Unless the training sample size is very small,[41] a best practice is always to use balanced stratified samples where there are an equal number of non-reference and reference target observations. With such balanced samples, the independent samples t value is not dependent upon the variance estimate calculation, as independent samples Student's t and Welch's t values are identical and they are also approximately identical to the t values based upon the Pearson correlations. In fact, balanced stratified sampling has become a standard approach in logistic regression modeling where intercepts are corrected at the scoring stage to reflect actual estimated imbalance in the relative proportion of binary outcome responses. So the best RELR practice of sample balancing accords with this already existing standard approach in logistic regression modeling.

Actual positive and negative errors always cancel each other to some extent, but this net error that results from the difference between the estimated positive and negative errors is the quantity that can be estimated through the logistic distribution in RELR. These positive and negative errors are assumed to be random variables, so there should be no greater probability of positive versus negative error for a given set of features. So, the second assumption assumes that the probability of positive and negative error are equal across features in the training sample used to build the model.

The third assumption relates to how RELR defines features. In RELR, an input variable is standardized with a mean of 0 and a standard deviation of 1 to define the linear feature and then all higher order polynomial and interaction features are computed based upon this standardized linear feature and then also standardized. This produces very different nonlinear and interaction effects than what would be observed based upon raw variables that are not standardized. For example, because all RELR variables are defined on the same standardized scale, they are not subject to the marginality problems where predictions may change with interaction and nonlinear effects simply based upon the scaling when there is arbitrary relative scaling of features.[42] Additionally, with raw unstandardized variables, features of the same input variable based upon polynomial terms with odd exponents are almost perfectly correlated to features based upon even exponents. Yet, there is almost no correlation between the odd and even polynomial features for the same input variable with the standardized features used in RELR. Thus, the reason that odd and even polynomial features are grouped separately in RELR is that these standardized linear and cubic features are very highly correlated as are standardized quadratic and quartic features where Pearson's r is typically >0.9 in each group, whereas correlations across groups are typically <0.1. This correlation grouping that separates standardized odd powered and even powered polynomial features is already obvious with as few as 10 observations in the simulated data shown in Fig. 2.3(a)–(c).

The third assumption in RELR's error model takes the second assumption of equal probability of positive and negative errors and applies it separately to odd (linear and cubic) versus even (quadratic and quartic) polynomial features. The result is that there is no bias to have a greater positive versus negative probability of error in independent variable features with even powered exponents compared to terms with odd exponents. This third assumption needs to be imposed as a constraint in the RELR model because there is naturally a tendency for correlations and t values to be larger for standardized polynomial features with even powered exponents compared to odd powered exponents, and this can have a biasing effect on the error model which is based upon t values which reflect correlations to the dependent variable.

Bias across polynomial effects has been reported previously in simulated datasets that have multicollinear relationships between independent variables.[43] However, the precise pattern of bias will depend upon whether raw versus standardized features are used and how the standardization is performed. The pattern of bias in the expected $1/t$ error across RELR's standardized polynomial features can be inferred from Fig. 2.4(a) which shows how the t values vary across the four components (linear, quadratic, cubic, and quartic) in a binary logistic regression forward simulated model where the logit was composed of an equal mix of all four components. Quadratic and quartic features give larger magnitude t values than linear and cubic features. This difference is quite substantial being almost exactly two times larger on average, and very reliable across all 30 simulation trials with each trial composed of 1000 independent observations as shown in Fig. 2.4(a).

Figure 2.4(b) shows two corresponding RELR and logistic regression models generated with simulated features described in Fig. 2.4(a) and with a random error signal designed to ensure that complete or quasi-complete did not occur so maximum likelihood convergence would occur. The RELR models do have a slight bias toward linear and quadratic effects relative to cubic and quartic effects, and this bias is reversed and more substantial in standard logistic regression models. However, the RELR models are more stable than the standard logistic regression models.

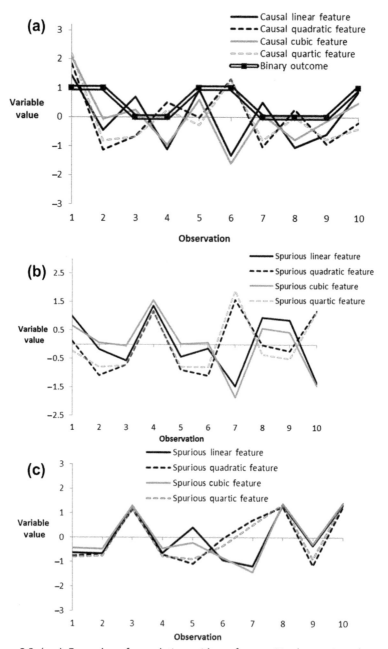

Figure 2.3 (a–c) Examples of correlation with as few as 10 observations between features of input variables in RELR's standardized variables. Standardized features with odd powered exponents (linear and cubic) are always highly correlated, whereas standardized features with even powered (quadratic and quartic) exponents are also

Without a nonbiased constraint to impose the third assumption, RELR's estimated expected error $1/t$ would be much higher on average for linear and cubic features. As a result, there would be a bias toward the selection of more false-positive effects in even power features relative to features that result from odd power exponents. RELR avoids this bias with the nonbiased constraint which forces the probability of positive and negative errors to be equally likely across features without any bias toward linear/cubic versus quartic/quadratic features.

With larger samples like the 1000 observations that were the basis of Fig. 2.4(a), another obvious pattern emerges. This is that there is a falloff in the sensitivity of t values with t values being more sensitive to linear features than cubic features and more sensitive to quadratic features than quartic features. The result is that RELR's expected error based upon inverse t values is biased to be lower with linear and quadratic effects than to cubic and quartic effects, so linear and quadratic effects would be more likely. This is especially true when the signs of the higher order cubic and quartic effects are reversed in simulations which only include causal linear and cubic features or only include causal quadratic and quartic features as shown in Fig. 2.5(a) and (b).

RELR does not attempt to correct for this particular bias that favors lower order linear and quadratic effects to have greater magnitude. Bias always will be a part of quantitative descriptions of reality because there is no perfect filter which has equal sensitivity to all aspects of the world. Instead filters necessarily must focus on some aspects of the world while diminishing others. The hope is that quantitative filters result in predictions and explanations that are similar to how human brains work. Human brains prefer to think in terms of simpler effects. That is, linear effects are preferred over cubic effects, and quadratic effects are preferred over quartic effects. This is because the human brain has a much easier time understanding simpler quantitative relationships. Even if a causal effect is more accurately described with cubic or quartic effects, most scientists would prefer to report a linear

---

highly correlated. These correlations average >0.9 in larger samples. However, odd power features do not correlate highly to even power features with correlations averaging below <0.1 in larger samples. An equal weighting of causal features in top panel and random noise signal that varied between −4 and 4 in the logit were used in a logistic regression forward model where any probability ≥0.5 = 1 and <0.5 = 0. Features in middle and bottom panels are spurious to this simulated binary outcome signal. See Fig. 2.4(a) for more details of simulation methods.

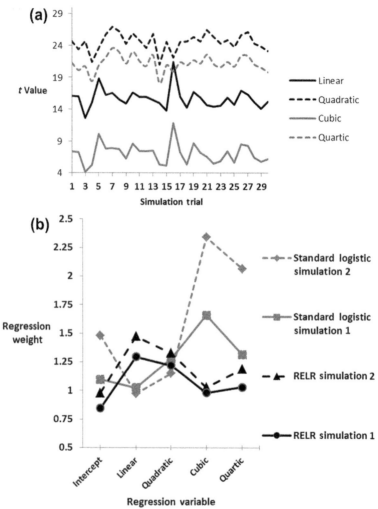

Figure 2.4 (a) Example of bias in *t* value patterns that is reliably observed across linear, quadratic, cubic and quartic *t* values created by employing a random number generator where signals varied between 0 and 1 to produce the linear feature *x* which was then mean centered and standardized. This is a representative sample of 30 trials of a simulation which consisted of an equal mix of each feature in a logistic regression forward simulated model where $y = \exp(x_s + (x_s^2)_s + (x_s^3)_s + (x_s^4)_s - \ln(0.29))/(1 + \exp(x_s + (x_s^2)_s + (x_s^3)_s + (x_s^4)_s - \ln(0.29)))$ and where each polynomial feature is not a raw variable but instead is a standardized variable with a mean of 0 and a standard deviation of 1 and the subscript s is used to remind readers of this standardization and its order as the nonlinear features are each created from the standardized linear feature x_s and then also standardized. In this simulation, a binary signal was created where any *y* value ≥ 0.5 was coded as 1 whereas any value < 0.5 was coded as zero across 1000

or quadratic effect if they fit data approximately almost as well. That is exactly what RELR does with its natural bias toward simpler effects.

In full models where parsimonious feature selection is not attempted, RELR is incapable of resolving linear versus cubic effects or quadratic versus quartic effects as separate effects often, but will have a bias toward placing greater magnitude weights on the lower order polynomial effects. In parsimonious feature selection models, this will mean that RELR will be much more likely to select linear and/or quadratic effects. This bias might be considered to be an advantage to RELR as it seems similar to what the human brain would be most likely to communicate on its own. Of course, if there is a very strong cubic or quartic effect that drowns out the lower order polynomial power effect, then RELR will observe such an effect.

While current software implementations only include polynomial terms up to the quartic effect, it is obvious that RELR could be applied in applications which employ much higher order polynomial features. Many important functions like sine and exponential functions can be approximated by a series of polynomial terms that includes higher order features in a Taylor series expansion over a defined interval. In fact, the Taylor series to approximate a sine function on the interval between $-\pi$ and π has as its first two terms x and $-x^3/6$ which is exactly what RELR gives back from the simulation in Fig. 2.5(a) as its expected standardized effects which are in fact the t values as reviewed above. This is because the means prior to standardization are roughly zero, whereas the standard deviations prior to standardization are roughly twice for the cubic versus the linear RELR

---

observations similar to the simulation described in Fig. 2.3(a)–(c) but with a much larger sample and no random error signal was used here. The intercept of $-\ln(0.29)$ was used because it gave approximately equal numbers of 0 and 1 signals. This binary signal was then correlated to each particular polynomial feature that was an independent variable feature and the t values for each feature and each simulation trial was determined. Clearly, even polynomial features tend to produce larger t values. (b) Comparison of RELR models versus standard logistic regression models in two simulations across 1000 observations created exactly as described for Fig. 2.4(a) except that a random number generated noise signal that varied with equal probability between -2 and 2 was now added to the forward logistic regression described in Fig. 2.4(a) to ensure maximum likelihood convergence in standard logistic regression. Clearly, the RELR models have greater weight in linear and quadratic and lower weight in cubic and quartic signals, whereas this bias is opposite for standard logistic regression. There is also substantially more instability (greater variability across the two simulation samples) in the Standard Logistic Regression weights compared to the RELR weights.

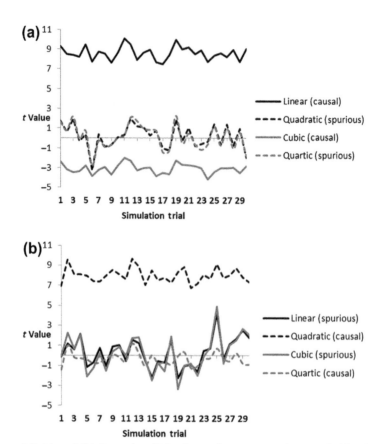

Figure 2.5 (a) and (b) Example of bias in t value patterns that is reliably observed across Linear, Quadratic, Cubic and Quartic t values in simulation which only included some features (causal) but estimated all features (noncausal features are labeled spurious). This simulation was also different from Fig. 2.4(a) in that the higher order cubic and quartic causal features were reversed in sign. The top figure is a representative sample of 30 trials of a simulation which consisted of an equal mix of each causal feature in a logistic regression forward simulated model where $y = \exp(x_s - (x_s^3)_s)/(1 + \exp(x_s - (x_s^3)_s))$ and where each polynomial feature is not a raw variable but instead is a standardized variable with a mean of 0 and a standard deviation of 1 and the subscript s is used to remind readers of this standardization and its order as the nonlinear features are each created from the standardized linear feature x_s and then also standardized. In this simulation, a binary signal was created where any y value ≥ 0.5 was coded as 1 whereas any value <0.5 was coded as zero across 1000 observations similar to the simulation described in Fig. 2.4(a) but also notice sign reversal of causal features in simulation. This binary signal was then correlated to each particular polynomial feature that was an independent variable feature and the t values for each feature and each simulation trial was determined. The bottom figure was produced similarly but only even polynomial terms were causal, so $y = \exp((x_s^2)_s - (x_s^4)_s)/(1 + \exp((x_s^2)_s - (x_s^4)_s))$. Notice that lower order causal features—linear and quadratic—clearly give larger t-value effects.

effect (Fig. 2.6(a) and (b)). On the other hand, t value effects are three times larger for positively weighted linear standardized versus negatively weighted cubic standardized variables (Fig. 2.5(a)). This implies that the $-x^3$ variable is divided by twice the linear term, whereas its expected effects once standardized are one-third the linear term. So the value in the denominator becomes 3!, which implies that the term becomes $-x^3/6$. So the effect of standardizing variables in RELR gives expected weights in the simulation of Fig. 2.5(a) which are exactly the first two terms of this Taylor series approximation to the sine function between $-\pi$ and π.

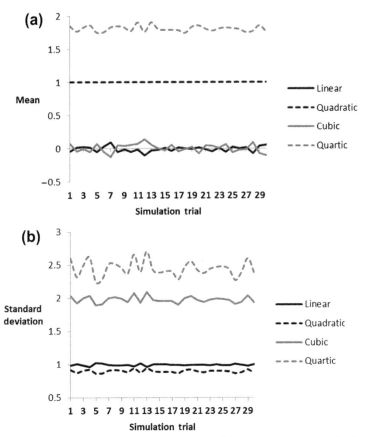

Figure 2.6 Means (a) and standard deviations (b) of features (prior to standardization) in a representative sample of simulation trials that employed features constructed in identical methods as used in Fig. 2.5(a) and (b). This particular simulation occurred across 1000 observations with balanced binary observations. Notice that means of linear and cubic features are roughly zero. Notice that standard deviations of cubic features are roughly double those of linear features.

In general, RELR does offer a method to fit a logistic regression model composed of polynomial features that could approximate other well-known functions in terms of expected effects through Taylor series. However, well-known functions like exponential, sine and cosine functions rapidly drop in terms of relative weighting of higher order polynomial features in Taylor series expansions. For example, the next term in the Taylor series sine function is $x^5/5!$, which is $1/120$th the weight of the linear x term. Because of this, most of the higher level terms might not be resolved in the presence of noise even with RELR's error reduction capabilities. So RELR's current software implementations that just include the first few polynomial terms might be close to a best approximation for these well-known analytic functions.

5. RELR AND THE JAYNES PRINCIPLE

RELR is designed to generate a most likely solution given the data and constraints. Unlike standard logistic regression, RELR takes into account uncertainty due to all sources of measurement error. Yet, the solutions that RELR produces cannot be said to be without error or bias, as clearly RELR's intercept and regression coefficients will be biased and will exhibit error. For example, intercept error and a greater weighting of lower order polynomial effects will exist in RELR. Yet, the most likely measurement error across all features will be expected to be estimated by the RELR error probability parameters. So, assuming that a minimal training sample is available so that intercept error also may be expected to be negligible, the RELR predictive probabilities may be expected to be relatively accurate. In fact, a fuller RELR model with many independent variable features does produce substantially reduced error in probability estimates. An example based upon real-world data is shown in Fig. 2.7.

At its core, RELR is simply based upon the Jaynes idea that the most probable inference is the maximum entropy solution subject to all known constraints. This simple data-driven approach to evidence-based reasoning has been successfully applied to scientific inference since Boltzmann. As with standard logistic regression, RELR's constrained maximum entropy solution is identical to the maximum likelihood solution. Indeed, RELR simply generates the most likely inference when error probability constraints related to uncertainty in measurements are added to the standard logistic regression model.

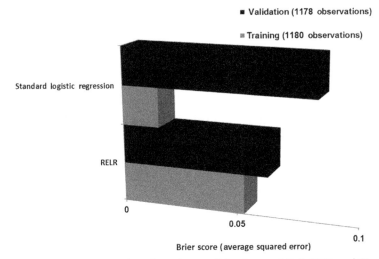

Figure 2.7 Accuracy comparison based upon Brier Score of Full RELR and Standard Logistic Regression models that have been previously reported[44] based upon public domain survey data on the 2004 US Presidential Election Weekend survey that was provided by Pew Research. The identical 176 features that included interaction and nonlinear features were included in both the RELR and the Standard Logistic Regression model. Notice how RELR shows substantially less overfitting error than Standard Logistic Regression as suggested by the Brier Scores that were substantially more accurate in the Validation sample. The methods used to produce this comparison are described in detail in Appendix A1.

Because of this ability to estimate error, RELR can model effects of higher level factors like schools or hospitals in correlated observations simply by including a dummy coded cluster indicator provided that the lowest observation level is composed of independent observations like persons. Any causal effects that exist at a higher cluster level such as related to specific schools or hospitals can be accounted for by the indicator variables, as RELR allows all levels of dummy coded indicator variables to be modeled without dropping a level to avoid multicollinearity error as must be done in standard logistic regression. This is one of the many advantages of RELR's ability to return stable regression coefficients due to its error modeling. Because of this error modeling, RELR has the ability to model uncertainty related to all measured and unmeasured constraints and return a most likely solution given that uncertainty. This goes beyond the Jaynes principle because the constraints that RELR uses to model error are not known with certainty but are instead only most likely constraints. Yet, RELR and its error modeling are still consistent with the overall Jaynes philosophy which is to arrive at a most likely inference.

In fact, RELR's explicit feature selection also can select parsimonious features that are not known with certainty in a causal sense, but are at least interpretable and consistent with causality until experimental data suggest otherwise. In addition, RELR's implicit feature selection can select a large number of features which are unlikely to be interpretable because of the complexity, but which are likely to generate a reasonable prediction provided that the predictive environment is stable. Alternatively, with a fixed set of features, RELR's sequential learning of longitudinal observations can detect the difference between a stable predictive environment versus an unstable and changing predictive environment and compute maximum probability weight changes accordingly. RELR's online learning has "memory" for previous observation episodes, which is not possible in standard logistic regression. Thus, RELR can model longitudinal data without violating independent observation assumptions. These maximum probability learning and memory abilities are reviewed in the next chapter.

CHAPTER 3

Probability Learning and Memory

> *"The probability of any event is the ratio between the values at which an expectation depending on the happening of the event ought to be computed, and the chance of the thing expected upon its happening."*
> **Reverend Thomas Bayes, An Essay Towards Solving a Problem in the Doctrine of Chances, 1763.**

Contents

This very cryptic passage from Thomas Bayes is credited with defining a view that probability is a measure of expectation or degree of belief. Yet, it was not until twentieth-century statistician R.A. Fisher developed the maximum likelihood method which views probability to represent the long-run frequencies of real events that the term Bayesian began to be used as an alternative to Fisher's frequentist view of probability.[1] Bayesians can be divided between those who believe that probability is subjective and can differ between persons no matter the evidence versus those who believe that probability is objective so given the same evidence rational people or machines will come to the same probability expectations.[2] Jaynes was an objective Bayesian, as his maximum entropy principle was designed to formalize how probabilities could be defined completely based upon known evidence or reasonable assumptions. Somewhat related to the frequentist view of probability is the view that probabilities reflect propensities or tendencies inherent in nature that allow predictions to be made. The real divide then between popular notions of probability is along the lines of whether probabilities are seen to be epistemic or an aspect of one's belief structure as Bayesians believe, or whether probabilities are instead ontic which means that they relate to an aspect of the real world.[3]

When viewed from the perspective of what defines learning and memory in the brain, this divide over the definition of probability seems very artificial. Whether conscious or unconscious, prior beliefs and memories in the brain are caused by a mechanism that is based upon previous learning. Granted, not all of the brain's memory structure would be determined by an individual's life history, as some of it may be instinctual and determined by the evolutionary history of the species. In any case, the mechanism that gives rise to memory is a learning mechanism based upon historical ontic events. The brain then may be viewed as a genetically determined biological machine that continually learns and revises these prior probabilities or memories through incremental learning. Different individuals may have different subjective beliefs when given the same evidence because they have different learning histories with differing ontic events. Yet, individuals with the very same learning histories should come to the very same probability expectations. And, in cases where learning observations can be continually repeated and independently replicated where every attempt is made to control the effect of prior beliefs like in controlled scientific experiments, it should be possible to demonstrate objectively that one given scenario is most probable. Otherwise, any possibility of ever reaching any scientific consensus on what is a most probable prediction or explanation would be impossible, which is clearly not the case as documented in the history of science. Can this most likely prediction or explanation be confusing and even incorrect? Yes and this has often happened in the history of science, but such correction is itself a scientific process which relies on new data and new most likely interpretation in what may be essentially a Bayesian incremental learning process.

1. BAYESIAN ONLINE LEARNING AND MEMORY

Online or incremental learning is driven by sequential learning episodes. This would be appropriate for near real-time learning applications. Online learning is different from batch learning which uses samples based upon entire historical periods in its learning. RELR's online learning is based upon the basic Bayesian notion that expectation based upon historical memories influences ultimate judgments of probability. This Bayes' rule can be expressed as

$$p(y|x) = p(x|y)p(y)/p(x) \tag{3.1}$$

which can be read as the posterior conditional probability of y given x or $p(y|x)$ equals the likelihood of x given that y happened or $p(x|y)$ multiplied by the ratio of the prior probability of y to the probability of x or $p(y)/p(x)$.

More simply, this expression tells us how to form probability expressions to update prior beliefs or memories when new information is learned so that posterior beliefs reflect both remembered and newly learned information.

This Bayesian incremental learning is easily understood in the so-called Monte Hall problem that is based upon the game show *Let's Make a Deal*. You are told that a Million Dollar Grand Prize is behind one of three doors, so you choose the first door. Then, you are told that it is not behind the second door and you may revise your initial choice. Your initial choice had only a 1 out of 3 chance of being correct. If you revise that choice to the third door, you now have a 1 out of 2 chance because only the first and third doors are left as possibilities. It would be wise to choose the third door, but many very intelligent humans fail to make this rational choice. As the Tversky and Kahneman research consistently has shown,[4] the reason appears to be that human decisions are very much affected by other factors unrelated to the actual information given. Maybe some people tend to anchor their decisions to their initial thought[5] possibly because of some genetic instinctual survival reason for human decisions to be so anchored. Or, maybe some people have explicit or implicit memories in their life history that cause them not to trust authority figures, so they consciously or unconsciously believe that the game show host is trying to trick them. If somehow all the genetic and environmental historical information that determines a person's subjective decision could be measured, then it could be that the decision process is actually a Bayesian process given the prior beliefs/memories and the new information that is learned. In any case, the end result is that the added prior subjective memory information that humans tend to use in such decisions often leads to less probable inferences than would be afforded by Bayesian learning simply based upon what is given as present and objective data.

The maximum entropy subject to constraints mechanism only generally yields most probable inferences given an initial or naïve state when the prior probability of all observed outcomes can be considered to be equally probable without seeing any data. Once outcome data become known as new information to be learned, a reasonable Bayesian approach would use the old posterior outcome probabilities as new prior outcome probabilities for this learning of new outcomes. So, with new outcome observations, previous posterior probabilities become the new prior probabilities and the most probable inference is that which is closest to these prior probabilities. In a logistic regression which aims to compute a posterior probability distribution **p** given a prior probability distribution **q** and associated constraints, this can be formalized as a process that minimizes the following expression known as the Kullback–Leibler or KL divergence subject to all associated constraints:

$$D(\mathbf{p}, \mathbf{q}) = \sum_{i=1}^{N} \sum_{j=1}^{C} p(i,j) \ln(p(i,j)/q(i,j)) \tag{3.2}$$

where i indexes the N observed outcomes and j indexes the C outcome categories just as in as in the discrete Shannon maximum entropy definition of the last chapter.

In real-world implementation, both **p** and **q** can be assumed to be everywhere positive here where they represent nonzero probabilities. In the special case where the prior probability distribution **q** is equal across all ith and jth conditions, the minimum KL divergence is identical to the Shannon maximum entropy. Otherwise, its minimum value will be different from the Shannon's standard information theory maximum entropy value because of the unequal prior probability distribution **q**. The KL divergence can be thought to be a measure of the amount of information that is gained when a posterior distribution is updated with new known constraint information compared with the prior distribution. When the historical observation probability distribution **p** is computed to match the prior distribution **q** perfectly subject to all constraints, there is no information gained. Otherwise, information is gained.

The KL divergence is not symmetric in the sense that viewing a probability distribution for one set of observations as the prior distribution **q** when updating a second distribution **p** based upon a different set of observations will not generally yield the same result as when these relationships are reversed.[6] Hence, a prior distribution **q** that is used in the minimal KL divergence usually has meaning as a probability distribution based upon earlier events in time when applied in RELR. Such a meaning is present in longitudinal data composed of panels or cohorts in sequential online samples of observations. As reviewed in Appendix A4, RELR's sequential online learning even allows time-varying effects like seasonal effects to be learned through sequential cross-sectional samples.

This probability distribution **q** based upon earlier events in time also can have meaning as a prior belief distribution based upon memories that can be updated with new information, as there are unique prior weights for each regression effect that is the basis of the prior distribution **q** as reviewed in

Appendix A4. This meaning has cognitive interpretation as the brain seems to update its implicit and explicit memories with new small incremental learning samples. That is, both historical prior memory weights and new data seem to exert influence on the posterior beliefs in the brain. A RELR online sequential learning model of square perception that shows the same Kanizsa illusion effect seen in humans based upon strong prior beliefs is presented in Chapter 6.

Maximum entropy estimation subject to appropriate constraints yields identical parameters as maximum likelihood logistic regression estimation as reviewed in the last chapter. More generally, maximum likelihood estimation gives equivalent solutions as the minimal Kullback–Leibler divergence of a hypothesized prior distribution **q** with prior probabilities that are not everywhere equal relative to a posterior probability distribution **p**.[7,8] In logistic regression, minimizing the KL divergence subject to constraints along with its equivalent maximum likelihood counterpart represents a Bayesian online learning rule because the new posterior probabilities can be expressed in the familiar Bayesian form that a posterior probability is proportional to a prior probability times a likelihood given by the data. In binary logistic regression, this becomes

$$p(i,j=1) \propto q(i,j=1) \frac{e^{\Delta\alpha+\sum_{r=1}^{M}\Delta\beta(r)x(i,r)}}{1+e^{\Delta\alpha+\sum_{r=1}^{M}\Delta\beta(r)x(i,r)}} \quad \text{for } i=1 \text{ to } N, \qquad (3.3a)$$

$$p(i,j=2) \propto q(i,j=2) \frac{1}{1+e^{\Delta\alpha+\sum_{r=1}^{M}\Delta\beta(r)x(i,r)}} \quad \text{for } i=1 \text{ to } N. \qquad (3.3b)$$

where the argument to the exponent is defined based upon regression weights and independent variable features just as in Eqns (2.5a) and (2.5b) except we need to remember that the update regression weights $\Delta\alpha$ and $\Delta\beta(r)$ for $r=1$ to M parameters now reflect a change in the parameters that were used to compute previous posterior probability distribution which is now the prior probability distribution **q**. This is only a proportional relationship because these expressions on the right need to be normalized so that they sum to 1 across the $j=1$ and $j=2$ binary categories for each ith observation so that they have the proper form of probabilities. This normalization is shown in Appendix A4. This allows a separation of prior regression weights from update regression weights, which in turn sum to give the posterior RELR regression weights which are the basis of the posterior distribution **p**.

In RELR's online learning based upon the KL divergence, only the observation posterior probability distribution **p** is determined in terms of a prior distribution **q** in the updated online model (Eqns (3.3a) and (3.3b)). The error probability distribution **w** is not updated using any prior empirical distribution ever. This is because error is expected to vary randomly across sequential learning episodes, so any estimates of error from previous episodes should not influence online update learning. Because of this distinction, RELR's online learning of an observation posterior probability distribution **p** can be strongly influenced by a prior distribution **q** when the new learning episode gives update parameters that are different from those that are the basis of **q**, whereas RELR's error probability distribution **w** is never influenced by a prior distribution. So in the case of RELR's online learning, the posterior distribution **p** has a memory for historical observations that are the basis of the prior distribution **q** when models are updated in new online learning episodes in longitudinal data like panel data that are composed of sequential cross-sectional samples.

The prior distribution **q** will influence the posterior distribution **p** in many applications including RELR, along with many applications that are not logistic regression estimations such as imaging applications. However, surprisingly the prior distribution **q** does not influence the values of the posterior distribution **p** in standard logistic regression. That is, in standard logistic regression, the posterior probabilities $p(i,j = 1)$ and $p(i,j = 2)$ in the left-hand side of Eqns (3.3a) and (3.3b) will not be affected by the form of the prior distribution **q**. Thus, there is no memory in the posterior online distribution **p** in terms of the historical data used to form the prior distribution **q** in standard logistic regression. In fact, the same posterior distribution solution **p** is found whether or not a prior distribution **q** is used, as not using a prior distribution is equivalent to using an everywhere equal prior distribution **q**. The reason is that standard logistic regression solutions are entirely determined by constraints that force binary probabilities for each outcome observation, and these solutions must also honor the data constraints given by the independent variable features without any allowance for errors-in-variables. So standard logistic regression is like the chameleon lizard that changes its color perfectly to adapt to each new situation. In this case, it is the posterior solution **p** which is changed to adapt perfectly to each new training sample of data without any memory influence from its prior distribution **q**.

The reason that RELR is different from standard logistic regression is that RELR estimates an error probability distribution **w** in its modeling which allows errors-in-variables that determine that regression coefficients across features that are in its solutions can be somewhat independent from the independent variable feature constraints. This independence also effectively creates some sensitivity in RELR's regression coefficient estimates across features to the prior information **q** that is used to update RELR's posterior probability distribution **p**, whereas this sensitivity is not seen in standard logistic regression. These differences between RELR and standard logistic regression are exemplified in regression weights obtained from simulations depicted in Fig. 3.1.

The simulations in Fig. 3.1 show both standard logistic regression and RELR regression weights produced both with and without a prior distribution at the sample size of 100. The prior distribution used in all cases was built with RELR with an independent sample size of 500. Besides the much greater overfitting error in standard logistic regression as exemplified by the much larger magnitude coefficients, it is also seen that the standard logistic regression weights are identical whether or not a prior distribution was used. On the other hand, the RELR regression weights do depend upon whether a prior was used at these sample sizes of 100. Not shown in this figure is that

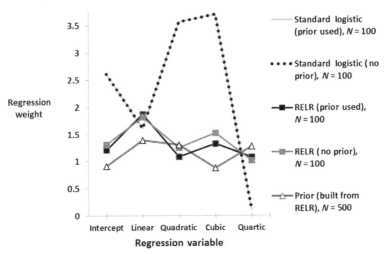

Figure 3.1 The effect of a prior distribution on standard logistic regression and RELR regression weights on a subsample of 100 observations in a simulation that was conducted exactly as described in Fig. 2.4(b). The Prior (Built from RELR), $N = 500$ was used as a prior for both the Standard Logistic (Prior Used), $N = 100$, and RELR (Prior Used), $N = 100$ condition. No prior was used in the other two conditions as noted. RELR posterior weights are sensitive to the whether or not a prior was used, whereas standard logistic weights are identical in each case.

at larger sample sizes in this simulation, RELR's solutions will depend much less on whether a prior distribution is used, so that at $N = 500$ the Prior Used and No Prior Used conditions for RELR are much more similar than at the $N = 100$ condition. However, the sample size at which RELR shows similar posterior solutions that depend less upon the prior is determined by data characteristics such as the number of features that are used in model, along with the degree to which the environment from which the data are sampled is stable. That is, RELR only will show posterior solutions that depend less on the prior in very stable environments that do not change substantially with new samples of sequential data.

So RELR's online learning has a memory in the sense that the posterior probability distribution will be affected by a prior probability distribution that is based upon historical observations. This allows RELR not only to detect environmental stability provided that its sample is large enough but also to detect environmental changes and instability across longitudinal data. In such sequential data, each new sample in time simply needs to be updated using the previous RELR posterior probability distribution $\mathbf{p}$, as this previous probability distribution $\mathbf{p}$ becomes the new prior probability distribution $\mathbf{q}$ in the minimal KL divergence. Because RELR can update its longitudinal observations in sequential online samples, all RELR models can be based upon the most current sample data where all observations can be assumed to be independent, and the prior probability distribution $\mathbf{q}$ can accumulate from previous samples and handle all dependencies that exist in correlated longitudinal data. The RELR historical memory in the form of the prior distribution $\mathbf{q}$ will decay eventually to zero with new online data that are changing in an unstable environment and so do not fit the historical data prior probability distribution $\mathbf{q}$, but the rate of any such memory decay will be highly data dependent.

The simulation example in Fig. 3.1 is controlled in that both samples are generated by exactly the same data generating instrument and can be assumed to be completely identical except for sampling error. This simulation is similar to what would be expected in longitudinal samples where all observations can be assumed to be independent from one another over time in the sense that an earlier observation does not cause a later observation but instead the observed dependencies or correlations over time are caused by the independent variable features. In such longitudinal data, RELR's online updating based upon the minimal KL divergence method will give similar results as batch learning across an entire stable historical period given a large enough sample at each update. However this will not generally be the case.

As noted earlier, because the KL divergence is not symmetric, it is not generally appropriate for samples that are not prior to one another such as independent cross-sectional samples. The only exception is the very special condition in this simulation where the environment can be assumed to be perfectly stable and the update samples are large enough so that the online updating eventually approximates the batch learning condition. However, when observations in earlier longitudinal samples reflect direct or latent causes that exert effects in later samples, these temporal relationships need to be considered. So in this type of data generating mechanism where outcomes in earlier samples clearly can exert causal effects in later longitudinal samples, the prior distribution should represent probabilities of events that are prior in time compared with the events represented by the posterior distribution.

With judiciously chosen stable learning episodes, RELR's online learning is effective when all feature constraints are preselected and the same in each new online learning episode composed of independent observations. Yet, as noted above, in many applications with longitudinal data, earlier observations in time may reflect causal effects that drive later observations, so such observations cannot be considered to be independent across time. In these cases, independent variable features based upon the earlier outcome observations need to be entered as independent variable features in RELR in what biostatisticians call transitional models, and an example will be shown with low-birth weight pregnancy data later in this chapter. Yet, even in this case, online updating can be used in RELR, and it will allow independent variable features based upon past observations to be features that are no different in form from other features. The important advantage relative to standard logistic regression modeling is that each new completely independent sample of observations can be modeled all by itself, but all previous samples still will influence this model because they influence the prior distribution q. So there is no need for generalized estimating equations (GEE) or problematic assumptions concerning fixed or random effects modeling to handle correlated longitudinal observations.

The applications of RELR's online sequential learning would generally include any time-dependent or longitudinal data. However, one particularly interesting application is survival analysis because it has received a large amount of study from the perspective of logistic regression.[9,10] This methodology treats failure events like the clinical diagnosis of disease, as discrete dummy-coded events in time, with appropriate censoring that excludes an observation for all time periods after a failure event has occurred. In addition, repeated observations from the same person are treated as independent events. Still there are some rather nagging problems when binary logistic regression is applied. First, repeated observations from the same person are not independent over time.[11] Second, the logit response function produces regression coefficients that are biased by the discrete sampling interval between adjacent periods of time, although this problem can be corrected with the use of a complementary log–log response.[12] Yet, the problem related to the assumption of independent observations cannot be corrected in this way.[13]

In contrast, RELR's binary logistic regression can be applied to survival analysis and overcome problems related to both nonindependent observations and noncontinuous time measurements as long as a few special data preparation requirements are met. First the data need to be coded for usage with binary logistic regression including censoring as is typically done in binary logistic regression survival analysis.[14] Second, RELR should be applied to sequential samples of these binary coded independent observations so that its sequential online learning can work just like with any other longitudinal data. A big advantage of viewing survival analysis in terms of discrete binary failure-related events is that a proportional hazards assumption is not forced, as the model can have covariates that are time functions.[15] This allows the baseline hazard rate which also can be time varying to be adjusted by covariate features that are interactions with time as determined by data.[16] This avoidance of the proportional hazards assumption is detailed in Appendix A4 which details how RELR uses the logit link in survival analysis in similar ways as with other longitudinal data, but with nuances related to allowing time-dependent covariates. Because of the time-dependent covariates, RELR's regression coefficients will not be biased by the discrete sampling episode, as estimates of hazard rates can be adjusted in individual observations through a covariate feature that measures the duration of the sampling interval between repeated observations in continuous time measurements and interacts with other covariates.

Everything that has been discussed up to now has assumed that the independent variable features are known. However, much greater flexibility is required than always having to use preselected or preknown features. Much more often, important features that are the basis of effective prediction and explanation are not known with any reasonable certainty. Thus, general approaches to learn most probable features are needed.

2. MOST PROBABLE FEATURES

The Jaynes principle is based upon an assumption that the world is often uncertain and subject to the whims of probability. So at best a most probable judgment is possible given known constraints on this world because there are an infinite number of possible outcomes which fit historical data corresponding to the same known features. The methodology presented here simply extends that logic and suggests that the predictive and explanatory features that force constraints on data-driven inferences also usually are not known with certainty, but at least a methodology can be devised that allows a most probable set of features to be chosen.

Most probable predictive or explanatory features need to be learned in processes that adequately account for the uncertainty. These processes must apply to all possible sets of candidate features that would be observed empirically no matter how large that set may be. Yet, in high-dimension candidate feature sets in predictive and explanatory modeling, a dimension reduction step prior to feature selection will be useful. This is called feature or variable reduction. This is done to put the data in a more manageable form for more efficient and more rapid processing, but it is very important that the feature reduction process should not influence the final form of the model. That is, it is important that the prediction and explanation be the same whether or not the feature reduction was performed, so the only effect of feature reduction is to make the computational processing more rapid.

In high-dimension problems that typically require feature reduction, RELR yields regression weight parameters that are directly proportional to the t-value that describes how reliably a feature correlates to the target outcome. This relationship directly results from the algebra of the RELR error model because when the positive and negative error probabilities tend to become equal for all features, the RELR error model forces a direct proportionality between t-values and regression weights across all features within both the even and odd polynomial feature sets. (see Appendix A1; Eqn (A1.17)). This near equality of positive and negative error probabilities in all features tends to occur as the number of features is large compared with the number of observations. And, RELR is able to handle more features than observations no matter how many features because it adds two pseudo-observations for each feature corresponding to inferred positive and negative error events.[17]

So in all high-dimensional cases, a simple univariate t-value can be run as a screening feature reduction step prior to building a model. In this case, a truncated set of features can be chosen that has the largest t-value magnitudes, and this will be the same set that would have the largest magnitude regression weights if a full model across all features were computed. Hence, this RELR feature reduction allows the reduction of dimensionality into what can be characterized as the most likely set of candidate features. In RELR, the feature reduction is performed so that there is no bias in the rejection of odd-powered versus even-powered polynomial components, given such potential bias as detailed in the last chapter. With longitudinal data that are updated sequentially in successive learning samples, feature reduction may proceed on the new sample exactly as if there are no longitudinal data.

Yet, feature reduction is only the first step, as a modeler needs to select further from this most likely set of features those features that will generalize to other data samples in a most likely prediction or explanation. The second step is actually what is usually called variable or feature selection. In order for feature selection to work, some measure of the likelihood of a model given the data must be used to grade different possible selections and select that feature set with the highest grade. An obvious choice for a measure that reflects how likely a feature selection model is given the data would be a log likelihood measure.[18] Yet, in RELR there are actually different log likelihood measures to consider which do have quite different properties. This is suggested by the form of Eqn (2.9) which defined the total RELR log likelihood value which can be written as

$$RLL = OLL + ELL \qquad (3.4)$$

where RLL is the RELR total log likelihood, OLL is the observation log likelihood which is identical to the standard logistic regression log likelihood and ELL is the error log likelihood. A maximal RLL reflects the most probable model given a set of independent variable features. Because of this, it is reasonable to expect that this optimal RLL measure also could be useful to select a most probable set of independent variable features. RLL is different from the log likelihood in standard logistic regression in one important sense. Whereas the maximum log likelihood across all possible feature selection sets is always observed in standard logistic regression when the largest possible feature set is selected, this RLL value almost always reaches its maximum when a parsimonious feature set is selected. This is because each time that an independent variable feature is dropped in RELR, two pseudo-observations corresponding to the positive and negative error probability estimates for that feature are also dropped. The dropping of the pseudo-observations in RELR causes a larger value of *ELL* when the model is refit. For this reason, RELR will generate a larger *RLL* value for all refitted models corresponding to dropped features and pseudo-observations until the remaining feature set is so small that it underfits the data.

It is very easy to verify that a likelihood value which is simply the product of the probability of all independent observation and pseudo-observation events will reflect a more probable scenario with fewer events assuming that all observation data are fit equivalently. This is exactly analogous to how the joint probability of two fair coin flipping outcome events will be $0.5 \times 0.5 = 0.25$, so any outcome scenario with just two coin flipping events is more probable than scenarios involving more events like three coin flipping events where the joint probability is $0.5 \times 0.5 \times 0.5 = 0.125$. So by dropping pseudo-observations when it drops independent variables, the variable selection in RELR tends to generate a larger total RELR log likelihood value *RLL* or a more probable solution with fewer independent variables assuming that the observation data fit is held constant.

A better fit gives a less negative *OLL* value. Fewer independent variable features give fewer pseudo-observations and a less negative *ELL*. So *RLL* is maximal in the independent variable feature set which gives both a relatively good fit to the training data and which does not require a relatively large number of independent variable features. Because *RLL* contains the additional error log likelihood component *ELL* which has a more negative value with a larger number of independent variables and corresponding pseudo-observations, *RLL* does not have the same interpretation as the standard logistic regression log likelihood value. On the other hand, the *OLL* component which reflects the observation log likelihood has the same form and interpretation as the standard logistic regression log likelihood value in the sense of being a measure of the goodness of fit in the training data.

Therefore, there are actually two RELR log likelihood values to consider in terms of being able to guide optimal feature selection. First, there is the total RELR log likelihood value or *RLL* which represents the sum of two components: observation log likelihood known as *OLL* and the error log likelihood known as *ELL*. *RLL* is the most probable solution which gives both a good fit and which also returns a parsimonious model. Second, there is just the portion that is identical in form to standard logistic regression which reflects the observation log likelihood or *OLL*. *OLL* is the most probable solution in the sense of the maximal joint probability of training outcome events when the error model log likelihood is ignored. *OLL* tends to be maximal and least negative in solutions with a very good relative fit, but these are often very complex solutions that are not parsimonious, as in Fig. 2.7 in the last chapter where the RELR solution had 176 features.

So two RELR independent variable feature selection learning methods are possible where learning entirely is based upon training sample data. The choice of which one to use will depend upon the objective of the feature selection. If the goal is only to get a good prediction, then this can be achieved with the feature selection set which maximizes *OLL* across all possible selections that are considered. On the other hand, if the goal is to generate a solution that gives a relatively good prediction and which also selects relatively few independent variable features in what may be interpretable as an explanation, then this will be achieved with the feature selection set which maximizes *RLL*. Practical experience suggests that the validation sample fit will be usually better with a solution that maximizes *OLL* compared with *RLL* with very small training samples. However, with larger training samples, the parsimonious solution that maximizes *RLL* often will be at least very comparable or even superior. There is actually related evidence from data mining contests that a very parsimonious solution can outperform more complex solutions, as this happened in the 2001 KDD cup contest where a model which selected only three features actually won the contest.[19]

It also turns out that only two effective and useful feature selection methods are possible depending upon whether feed forward or feedback feature selection control is allowed and given that either the observation log likelihood *OLL* or the total RELR log likelihood *RLL* is used to guide feature selection. When the goal is pure prediction and there is no concern about the number of features selected, feed forward processing can be used like the previously described feature reduction process in that features can be selected based upon *t*-value magnitudes while honoring bias considerations.[20] In this process, all of the information required to compute one particular feature selection solution can be precomputed and run in parallel without requiring any information about any other feature selection solution. Only a last sequential stage is required that evaluates the observation log likelihood *OLL* of each feature set selection model and chooses that solution with the largest observation log likelihood as the best fitting such model to the training data.[21] This process is called Implicit RELR feature selection because it has similar properties as a neural implicit learning process in the sense that it is only concerned with a predictive model without an explanatory interpretation. On the other hand, when the goal is an interpretable explanation so there is concern about the number of features selected, the total RELR log likelihood *RLL* needs to be optimized across all possible independent variable feature sets. This process is called Explicit RELR because it does have similarity to neural explicit learning in that it does often return models that lend themselves to explanatory interpretation. However, Explicit RELR cannot be implemented entirely in parallel in general. These Implicit and Explicit RELR feature learning processes are completely automated mechanical machine learning processes and are detailed in the next two sections.

3. IMPLICIT RELR

Implicit RELR refers to any diffuse RELR model that does not result from parsimonious feature selection. So, Implicit RELR is actually a more generic term than just a feature selection process that may include, for example, a Full RELR model that includes all features.

In the Implicit RELR feature selection process, models are iteratively refit after each backward selection step based upon a univariate feature selection process which drops features based upon the magnitude of *t*-values that reflect the reliability of the feature's correlation to the target feature. The basic difference between the Implicit RELR feature selection compared with RELR's feature reduction is that Implicit RELR feature selection seeks a solution that maximizes the observation log likelihood *OLL* across a sequence of different feature selection sets that differ in terms of dropped features based upon *t*-value magnitudes, whereas RELR's feature reduction process is simply designed to reduce the total number of features using *t*-value magnitudes without ever running a model and computing *OLL*. Just like the RELR feature reduction process, each step in this Implicit RELR backward selection process drops one odd- and one even-powered polynomial feature simultaneously to ensure that there is no bias in dropped features. With high-dimension data, the constituents of the actual feature set that remains in Implicit RELR are usually not individually important because each feature receives relatively less weight because Implicit RELR usually selects a large number of features. This is unlike Explicit RELR where the parsimonious features that are selected are individually important (compare Fig. 3.2(a) with Fig. 3.2(c) or Fig. 3.2(d)). In high-dimension data, because there is substantial overlap and redundancy in selected features, different multicollinear features are almost always exchangeable with little effect on predictive performance. Thus, when Implicit RELR returns a large number of features in a Full model or feature selection solution, it is much more like a nonparametric model in the sense that the model's predictive accuracy does not depend heavily upon regression coefficient parameters from individual features.

Implicit RELR feature selection can be conceptualized as a form of ensemble learning that selects the weighting of feature selection models that gives the most probable fit to the training data in terms of the maximal observation log likelihood solution given the requirements in how the feed forward selection proceeds. Yet, the type of ensemble process in Implicit RELR is quite distinct from the normal meaning of this term in machine learning. In machine learning ensemble modeling, a set of features are employed to construct a model over and over again with different feature sets in each iteration and all are averaged together to give more stable weights than a model built from just one set of features.[22] This is very effective when there is adequate feature importance sampling to ensure convergence to a very accurate grand average model. The effect of such ensemble averaging is to arrive at feature weights in important features which are relatively free of error. Implicit RELR's feature selection is designed to achieve this same end, but does so directly through its error modeling and optimal *OLL* solution.

That is, the effect of the error model in RELR, which forces the probability of positive and negative errors to cancel without bias, is expected to have a similar effect as a well-constructed ensemble machine learning process which averages a large number of different submodels to achieve a grand average with much lower variability and error. For example, if logistic regression models were developed from a large number of random resamples of important features from a high-dimension set of features so that multi-collinear features would be unlikely to be sampled in the same set, it would be reasonable to expect the average of these feature weights across all models would converge to something approaching the univariate effect of each feature. That is, just like in RELR the expected error in the feature weight across samples would be equally likely to be positive and negative without bias in terms of even versus odd polynomial features. And, the effect of a given feature is ultimately estimated in such a grand average ensemble model while averaging out the effect of all other features which are not correlated because multicollinear features are unlikely to be sampled in a given sub-model. In addition, even if multicollinear features were selected in a given

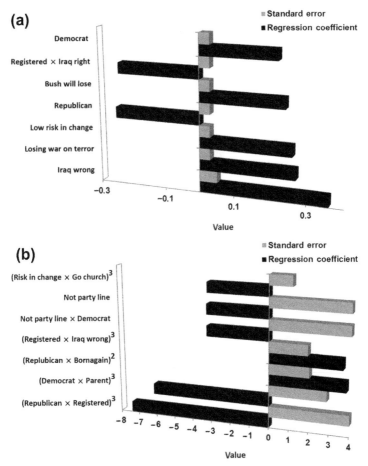

Figure 3.2 (a) and (b) Comparison of most important features in Implicit RELR model in Fig. 3.2(a) (top) with L2-Penalized Ridge Regression in Fig. 3.2(b) (bottom). Both models were based upon the identical 176 features. Notice how Implicit RELR's features have regression coefficients and standard errors of much lower magnitude. Also, notice how Implicit RELR's most important features are far simpler and linear without many interactions, whereas Ridge Regression is mostly composed of complex interaction and nonlinear effects in its most important features. These data are from the Pew US Presidential Election Weekend Survey in 2004, and the full details of these methods are in the Appendix A3. (c) and (d) Comparison of selected features in two Explicit RELR models based upon independent samples using the same 2004 Pew US Presidential Election Weekend Survey data used in Fig. 3.2(a) and (b). These samples had roughly 1180 respondents in each independent sample although the binary strata for target and non-target responses were only nearly and not perfectly balanced. The intercepts are not shown but both were approximately −0.04. Notice the stability of the regression coefficients, along with the face validity of the selected features. All regression coefficients have the same sign as correlation coefficients. The most important features in this model relating to the Iraq War and the War on Terror were generally thought to be the number one issues in this 2004 election.

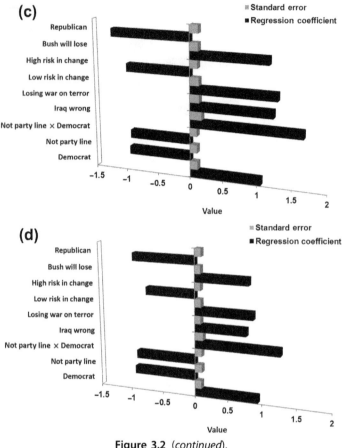

Figure 3.2 (*continued*).

sample of features, the probability of positive and negative error should be the same on average. So this error should cancel out on average. So the feature weights in the grand average should be similar to the univariate effects that would be obtained with uncorrelated, orthogonal features using standardized features just as in Implicit RELR built with a similar high-dimension set of features.

Therefore, given a large enough set of random resamples of important features from a high-dimension set of candidate features, a grand average ensemble model with regression weights that are proportional to univariate t-values like in Implicit RELR would not be surprising. In theory, the choice of the feature selection set with the largest OLL value in Implicit RELR feature selection might have a similar effect as importance sampling in ensemble machine learning theory,[23] as it ensures that those features which have the most important predictive effects are preferentially selected.

The *OLL* measure of a good fit in Implicit RELR could be better without precomputed steps based upon *t*-value magnitudes or without the error model constraints which determine that overfitting does not occur. Yet, an absolute maximal *OLL* value in the training sample across all possible solutions is not really the aim. Instead the objective is to get the largest stable *OLL* value with appropriate controls to ensure that this implicit learning process is completely automated and almost completely an embarrassingly parallel feed forward process that does not overfit substantially. The key here is the stability, as the error modeling constraints are necessary to ensure that this maximal *OLL* value will be stable and generalize well to samples other than the training sample. Under processing constraints in current software implementations that force the precomputed feature selection process to drop an even- and odd-powered feature at each step,[24] Implicit RELR feature selection is the optimal solution in terms of the fit defined by the RELR observation log likelihood across all such solutions. Unlike the standard ensemble learning methods in machine learning like Random Forests and related Stochastic Gradient Learning or Stacked Ensembles which are averages of many other elementary models, Implicit RELR feature selection is a direct computation of the solution which gives the largest *OLL* value under the feed forward processing and error modeling constraints.

Implicit RELR feature selection is interpretable as an average solution across a large number of features when no restriction is placed on solutions in terms of preferring parsimonious features so the only objective is to have the best observation log likelihood fit to the training data while honoring the error model and the other processing constraints. In the averaging process employed by other machine learning ensemble methods, the ideal would be that random perturbations would cancel across all features. The RELR error modeling essentially performs this same ensemble averaging process that forces random error to cancel out and not influence the solution. The end result is that Implicit RELR feature selection is a direct and automatic ensemble solution that does not require modelers to make decisions about how to weight aspects of the ensemble average and does not require time-consuming averaging across hundreds or even thousands of elementary solutions. Unlike machine learning methods such as super learners that compare many methods and select that with the best cross-validation loss score,[25] Implicit RELR also does not employ cross-validation or bootstrapping samples for its learning. Obviously, different cross-validation samples and/or loss functions may give different resulting definitions of a "best model" so RELR has an advantage that does not require users to make these choices.

The advantages of Implicit RELR feature selection are that it is a very fast and accurate automatic process that can be almost entirely implemented in parallel. While comparisons to standard predictive modeling methods consistently show that both Implicit and Explicit RELR are reasonably accurate as in Fig. 3.3(a) and (b), it needs to be reiterated that Implicit RELR feature selection is usually more accurate than Explicit RELR with very small training samples. The fact that Implicit RELR often returns regression coefficients that are approximately proportional to t-values as the number of features gets large is consistent with what is expected of accurate and stable methods for high-dimension data. For example Naïve Bayes often performs extremely well with high-dimension data and small sample sizes.[27] Naïve Bayes also yields regression coefficients that are proportional to t-values, although unlike RELR it does not model interactions. RELR comparisons to Naïve Bayes are reviewed in detail in the next chapter.

Implicit RELR does not select a large number of features in the rare cases with very few features with higher magnitude correlations, and in such low-dimension selections the regression coefficients are no longer proportional to t-values. Yet, even in lower dimension selections, feature selection based upon t-values is known to perform very well. For example, one recent study that compared 32 machine learning methods on accuracy, stability and interpretability in four different high-dimension genomic data sets found that simply selecting features based upon the top nine t-value magnitudes did the best overall compared with all other feature selection methods which included three different sophisticated ensemble aggregating methods, recursive feature elimination using support vector machines, LASSO, and Elastic Net.[28] For example, while Elastic Net and LASSO beat t-test selection slightly on one data set, they did not on others and t-tests were more stable and interpretable. This study has limitations in that although its selection was based upon t-value magnitudes, it only looked at a predefined number of features and a predefined small group of postselection modeling procedures including Naïve Bayes, SVM, Linear Discriminant Analysis, Nearest Centroid, and K Nearest Neighbors. Also, this study used the area under the curve (AUC) as its measure of accuracy which has known unreliability and validity problems.[29] However, the AUC anomaly problems

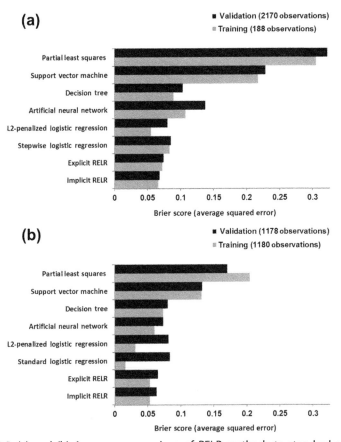

Figure 3.3 (a) and (b) Accuracy comparison of RELR methods to standard methods based upon Brier Score (black box—smaller training sample, gray box—larger training sample). In the bottom panel, note that both Implicit RELR and Standard Logistic Regression are the same as shown in Fig. 2.7. The other methods, including Explicit RELR, are now filled in for a wider comparison, as is data from the small training sample at the top. All models have been previously reported[26] based upon public domain data on the 2004 US Presidential Election Weekend Survey by Pew Research. Notice how Explicit RELR performs roughly the same as Implicit RELR in the bottom panel's validation based upon a larger training sample, and both show better performance than all other methods. The methods used to produce this comparison are described in detail in Appendix A3. (c) and (d) Accuracy comparison of RELR methods to standard methods based upon Classification Error Proportion (black box—smaller training sample, gray box—larger training sample). The p levels reported are all in comparison to Implicit RELR's validation performance. Note that L2-Penalized Logistic Regression used the validation sample error proportion for its learning, so this error proportion could be higher in samples not used to determine modeling parameters. All models are based upon same methodology as used in Fig. 3.3(a) and (b). Unlike the Brier Scores in Fig. 3.3(a) and (b) which were reliably more accurate in the RELR models in both the smaller and larger training samples, these classification error proportions are mostly only statistically better in Implicit RELR in the smaller training sample condition.

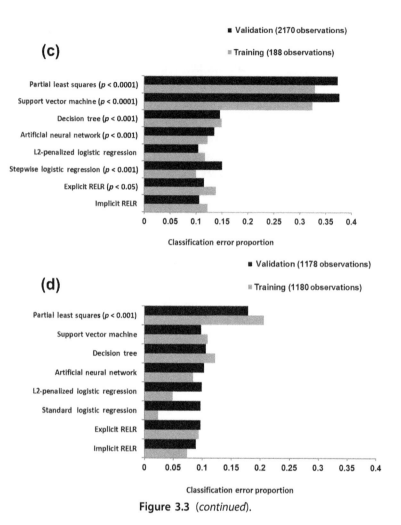

Figure 3.3 (*continued*).

would be less likely to appear in the very same form across multiple independent comparisons. So, these results do provide independent corroborative evidence that methods like Implicit RELR feature selection that return variable selections simply based upon univariate *t*-value magnitudes are quite competitive to traditional predictive modeling approaches including Elastic Net, LASSO, SVM, and ensemble modeling in high-dimension/small sample data.

Usually, more accurate machine learning methods only differ by a small percentage from one another, so accuracy is not the best determinant of usefulness in real-world application. Instead, along with stability and interpretability, the effort needed to achieve reasonable accuracy is a much better determinant given reasonable accuracy. In this regard, Implicit RELR feature selection is as automatic as can be without any arbitrary user-controlled aspects. That is, Implicit RELR feature selection does not require a user to predetermine the number of features to be selected based upon *t*-values, but instead it tells the user the number of features that optimizes the training sample observation log likelihood *OLL*. Methods that require a large amount of tweaking based upon independent validation sample results eventually may give more accurate solutions than Implicit RELR feature selection simply because Implicit RELR's solution is only the most likely fit given the training data and the feed forward constraints on how selection may proceed. Most data mining contests allow contestants unlimited submissions based upon tweaking and re-tweaking of models given feedback from validation sample results. Implicit RELR on the other hand is a one shot solution based entirely on the training data. No comparison has yet been made with Implicit RELR feature selection to other predictive modeling methods under conditions that allow accuracy from validation sample results to guide model building. But, Implicit RELR seems to perform well in apple-to-apple comparisons on validation sample accuracy where learning was based entirely on the training sample data without any feedback from even a small proportion of the validation sample.

While Implicit RELR usually does not allow interpretable models because of the complexity of its solutions, this is not to suggest that Implicit RELR always cannot reflect causal features. It is just that the features that are selected in Implicit RELR are usually too numerous or too complex to allow any understandable causal interpretation, although there always is the possibility that latent causes are manifested in the selected features. Whether Implicit RELR reflects causal or spurious features is not an issue when the predictive environment can be assumed to be stable like in a model that is formed based upon a cross-sectional sample and then is generalized to new cross-sectional samples that can be assumed not to have changed from the modeling sample. This is because features that are spuriously correlated to outcomes still can have good predictive accuracy when it can be assumed that the sample used to train a model is based upon similar feature values and combinations of feature values as the samples to which one wishes to generalize. Yet, when the predictive environment changes from the training sample, a model with spurious features can fail. A case in point is how a consumer's housing equity at the time of a refinance loan could be used as a good predictor of low risk in mortgage refinance loans in the United States until about 2007. Yet, when housing prices dropped precipitously, it became clear that this was actually a spurious feature which was often a very poor predictor of lack of loan default. So in a potentially changing predictive environment, it could be risky to use an Implicit RELR feature selection model because it is highly likely to select spurious features which just happen to have relatively large correlations to the outcome variable.

As another example of potential problems with purely predictive methods like Implicit RELR, suppose that there is a restaurant which tends to attract older adults on warmer days and younger adults on colder days because the older people tend not to venture out on colder days and because the younger people are at the beach on warmer days. This is an example of an interaction between weather temperature and age. If the historical data were only composed of data from warmer days, then any predictive model would erroneously predict that the restaurant's customer profile is primarily an older audience. This is because the training sample of data used to build a model is biased and does not include colder temperature values. This sample selection bias and the corresponding predictive problems when the weather turns colder might be obvious to anyone who understands how the model makes predictions, but it may not be obvious in models that are more difficult to interpret like an Implicit RELR feature selection which typically contains many redundant and overlapping features. Obviously, such a seasonality effect would be seen in an Implicit RELR model that included representative samples of consumers across all four seasons, so some of this weakness can be controlled through appropriate sampling but it is not always clear how to sample to avoid sample selection bias with other effects.

For example, a similar sample bias problem that missed an important interaction may have been the reason that some large US automakers did not realize that demand for sport utility vehicles would collapse when gas prices went up dramatically starting around 2007, but a similar dramatic shift in demand to small cars actually had happened previously in the very early 1980s when gas prices also spiked. In general, a pure prediction model like Implicit RELR which does not allow easy explanatory interpretation may be more likely to miss important interactions and have other sample bias feature selection problems.

Home loan defaults and automobile demand are longer term predictions. For this reason, it is hard to correct the model's poor decisions with sequential learning when it is realized too late that the original models have substantial future error due to spurious values and relationships. Thus, like other pure prediction methods, Implicit RELR may be most useful for short-term prediction situations or for situations where the predictive environment can be assumed to be very stable. Explicit RELR may be for any circumstance where is it expected that the predictive environment may change over time due to causal interactions or where underlying putative causal reasons for predicted outcomes are required for relatively safe longer term predictions, and where simulations could test effects of sudden changes in spurious predictors like home equity. This is because it will be much more obvious that a critical interaction is missing or a spurious relationship is present in Explicit RELR than in Implicit RELR.

4. EXPLICIT RELR

Although Explicit RELR selects far fewer features than Implicit RELR, there is no assurance that Explicit RELR's solution always will be simple enough to allow explanatory interpretation. In many cases, features that reflect latent causes may be selected which can be more difficult to interpret than features that reflect causes that are directly manifested in independent variable features. Other than returning relatively parsimonious solutions, Explicit RELR is similar to Implicit RELR feature selection in that it is a completely automated process and also tends to have accurate solutions with very little overfitting given a minimal sample size. The Explicit RELR process is a backward selection process that begins with a set of features that are potential candidate features. When there are a large number of original features in relation to the number of observations, RELR feature reduction based upon the magnitude of t-values is applied as the premodeling step.

In higher dimension RELR models built from multicollinear features, all χ^2 estimates for odd polynomial features are roughly similar in value as are all such estimates for even polynomial features due to the error modeling constraints. This effect is usually observed even in relatively low-dimension models built from relatively small training samples. Examples of RELR models where all features have similar χ^2 values because they have similar regression coefficient and standard error magnitudes can be observed in Fig. 3.2(a) and (b). The relationship between χ^2 values, regression coefficients, and standard errors for the estimated feature effects can be understood through the Wald relationship for the χ^2 in logistic regression. The

Wald relationship characterizes the large sample behavior of the χ^2 measure in standard logistic regression where

$$\chi_r^2 = \left(\frac{\beta_r}{se_r}\right)^2 \tag{3.5a}$$

for all $r = 1$ to M features where β_r is the regression coefficient for that feature and se_r is the standard error of the regression coefficient. Yet, RELR mimics standard logistic regression's large sample reliable behavior in small samples, as Wald approximations can be used to compute the χ^2 effects in RELR in small samples and give very similar values as the Likelihood Ratio test. The Likelihood Ratio test assesses the effect of removing one parameter which is approximated as χ^2.[30] Recall that in RELR the regression coefficients are always standardized regression coefficients, as all features are always standardized with a standard deviation of 1 and a mean of 0. Hence, when selected RELR features have similar χ^2 effects in high-dimension and/or small sample size situations, this necessarily implies that the standard error of the regression coefficient is similarly almost proportional to the standardized regression coefficients across features given the Wald relationship. In other words, features with small magnitude regression coefficients will have small standard errors in regression coefficient estimates, whereas features with larger magnitude regression coefficients will have larger standard errors in these estimates. The standard errors here are an estimate of the standard deviation in the sampling distribution of β values for a given feature across independent samples. Because β is expected to be proportional to univariate t-values in high-dimension and/or small sample models in RELR, it also makes sense that there also should be a correlation between β and its standard error and that the magnitude of χ^2 should be similar across features in these cases.[31]

So unlike the univariate t-value magnitude, the magnitude of the χ^2 measure shows substantially low variability in high-dimension and/or small samples across all features in RELR. Because of this, χ^2 provides relatively poor feature importance discrimination information. In fact, as shown below Explicit RELR's feature importance measure is actually based upon the product of χ (the square root of χ^2) and the regression coefficient β for each given feature. Hence, given that t and β are highly correlated in higher dimension RELR solutions and given that χ^2 values are roughly constant across these same high-dimension features, reducing a high-dimension feature set by using the univariate t-value magnitudes excludes those features that also would be deemed relatively unimportant in the Explicit

RELR selection. At least, this can be expected to be true until the selected feature set is of low enough dimension that the χ^2 values begin to show more variability across features. Yet, care always must be taken to start the Explicit RELR solution with a large enough set of candidate features so that χ^2 shows this low variability across features and the feature selection is not biased by where it starts.[32]

Once an initial candidate set is identified based upon t-value magnitudes, Explicit RELR simply drops one feature at a time from the larger candidate set and iteratively refits the model at each step. Among all such models, the model with the largest RLL value is chosen as the best model. The feature which is dropped at each step is the least important feature, where importance is defined in terms of how sensitive the χ^2 measure associated with a feature effect is to a change in the stability s of the regression coefficient where stability or s is the inverse standard error $1/se$ of the regression coefficient for the feature. That is we seek the derivative of χ^2 with respect to the stability s of the regression coefficient for the same effect, and view the derivative with the least magnitude to represent the least important feature. The stability s is an estimate of the extent to which a regression coefficient will match its population value and so be similar across independent samples. Like standard logistic regression, RELR's regression coefficient stability increases with an increasing sample size. But unlike standard logistic regression where regression coefficient stability usually becomes much lower with larger numbers of features, RELR's regression coefficient stability actually becomes much higher with larger numbers of features.[33]

The χ^2 value for an effect measures how reliably different from zero a regression coefficient is because it reflects how much the RELR log likelihood changes with the zeroing of that parameter. On the other hand, stability measures how representative the regression coefficient resulting from the training sample will be of a value seen in a larger population. It is possible to have a large χ^2 effect associated with a very small magnitude regression coefficient or vice versa depending upon its stability. So, both of these measures should be involved in a measure of feature importance. A very important feature with a very large χ^2 value and with regression coefficient stability that approaches infinity would be very sensitive to even a small change in stability. On the other hand, a very unimportant feature with a very small χ^2 value with regression coefficient stability that approaches zero would change very little with a small change in stability. Thus, the magnitude of the derivative of χ^2 with respect to s or stability has face validity as a measure of feature importance. With an assumption that the

Wald χ^2 estimate in RELR approximates its large sample value which happens in RELR with even very small sample sizes, the magnitude of this derivative can be derived from the Wald χ^2 formula in logistic regression as shown in the appendix. This derivation results from the chain rule in elementary calculus. With the Leibniz notation the magnitude of this derivative can be expressed as

$$\left| \frac{d\chi_r^2}{d(s_r)} \right| = |2\chi_r \beta_r| \tag{3.5b}$$

for each of the $r = 1$ to M features that remain as candidates in the Explicit RELR backward selection at a given iteration. Thus, the magnitude of the derivative of χ^2 for a given effect with respect to the stability s of the regression coefficient for the same effect is Explicit RELR's measure of feature importance. As noted above, χ^2 values show very low variability across RELR features, so the ranking of χ^2 values is a very poor measure of the relative importance of a feature. This is especially true when the relative importance ranking must be stable across data samples so that different features are not randomly selected across different samples as is required in stable feature selection. Yet, there can be substantial variability across samples in the selected variables when χ^2 is used to rank order the importance of features in Explicit RELR. A particular set of features selected based upon χ^2 might even fit the Explicit RELR *RLL* objective very well in the training data, but it still will be unlikely to generalize well to other samples because it is not stable. On the other hand, RELR's feature importance measure takes into account stability, and so it is much more likely to give similar relative importance rankings across models built from independent samples.[34]

The reason that Explicit RELR drops the least important feature in backward selection is suggested by the form of its *RLL* objective function. In particular, because its *ELL* component is composed of error probability estimates for features that are reflected as pseudo-observations that are always dropped in tandem with independent variable features, then it is clear that the more independent variable features that can be dropped, the larger will be the *ELL* value. So, assuming that the other component *OLL* is equivalent or at least little changed across different possible solutions, it is clear that the solution that has the fewest independent variable features will also have the largest total RELR log likelihood also known as *RLL*. So, the goal of a method that maximizes the total RELR log likelihood *RLL* must be to try to drop as many independent variable features as possible in a solution that still gives a good fit in terms of the observation log likelihood OLL. The χ^2 measure for a given independent variable feature is approximately minus two times the ratio of the current log likelihood RLL to that which is found when the feature is zeroed. So by zeroing and thus removing the feature effect with the smallest rate of change in χ^2 per change in the stability s of the regression coefficient at each step, a slowest and most stable possible trajectory is discovered in this ascent to the maximum RLL where we also drop each feature's pseudo-observations at each step. This slowest and most stable possible trajectory ensures that the maximum possible steps are taken before the maximal RLL across all steps is found. In this way, what is likely to be the maximal RLL solution at least in out-of-sample data is also discovered.

Figure 3.4(a) shows trajectories of the total RELR log likelihood RLL, the error log likelihood ELL and the observation log likelihood OLL in Explicit RELR feature selection in a model based upon a Low–Birth Weight data set, along with corresponding feature weights at each iteration step. This Low–Birth Weight data set is a widely used standard logistic regression data set developed by Hosmer and Lemeshow and available online.[35] The outcome variable is whether a low-birth weight pregnancy was observed or not. A balanced stratified cross-sectional sample of 56 observations was randomly selected from this larger data set to result in the training sample here. This sample was selected from the next to last cross-sectional sample in that data set, as the last cross-sectional sample was used for out-of-time testing of a putative causal effect as reported in the next chapter. The toy model presented here also does not contain any prior offsets, nonlinear effects or interaction effects[36] just so to be very simple. An intercept was also excluded here. The rationale for excluding the intercept in this case will be in Appendix A7 within the context of how a sample that is stratified to be composed of an equal balance of target and nontarget outcomes as in this case has a reasonable prior value in RELR for the intercept of zero. In general, RELR will perform intercept correction as a postmodeling step to correct for such stratification.[37]

As apparent in Fig. 3.4(b) early in the Explicit RELR trajectory features that are dropped have almost no effect on future feature weights, so this is a very stable process. Only later in the process when important highly collinear features are dropped are there large changes in remaining features' weights. For example, the PREVLOWTOTAL feature reflects the total number of previous low-birth weight pregnancies, whereas the LASTLOW feature reflects whether the last pregnancy had a low birth weight. These two features were correlated with one another roughly 0.9

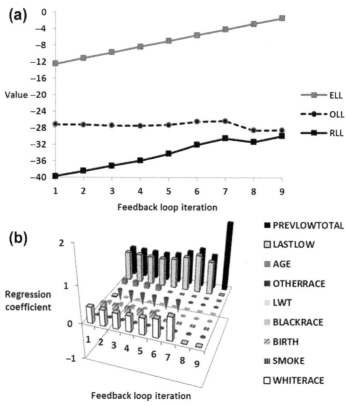

(a)

Value

ELL
OLL
RLL

Feedback loop iteration

(b)

Regression coefficient

Feedback loop iteration

■ PREVLOWTOTAL
▨ LASTLOW
▨ AGE
■ OTHERRACE
▨ LWT
▨ BLACKRACE
▨ BIRTH
▥ SMOKE
☐ WHITERACE

Figure 3.4 (a) and (b) An Explicit RELR model to predict low birth weight in a human sample. Top panel shows the trajectory of the Explicit RELR computations in terms of the *RLL* (RELR log likelihood), *ELL* (error log likelihood) and *OLL* (observation log likelihood) values across nine feedback iterations in a sample of 56 observations equally split in target and nontarget outcomes. Upside down cones show negative coefficients. Bottom panel shows the RELR regression coefficients across these same nine feedback iterations. (c) and (d) This is a model like Explicit RELR in every way except that χ^2 is used to determine the least important feature in backward selection instead of the derivative of χ^2. Otherwise, everything else is exactly as in Fig. 3.4(a) and (b) where upside down cones show negative coefficients.

across observations in this training sample and both were kept in the Explicit RELR feature set until the very last iteration. The OTH-ERRACE and WHITERACE were also kept in the feature set until late in the process. Note that because RELR is able to handle extreme multicollinearity, it dummy codes all levels of a categorical variable like Race and actually models each as a separate standardized variable. This is unlike how standard logistic regression typically handles dummy coded

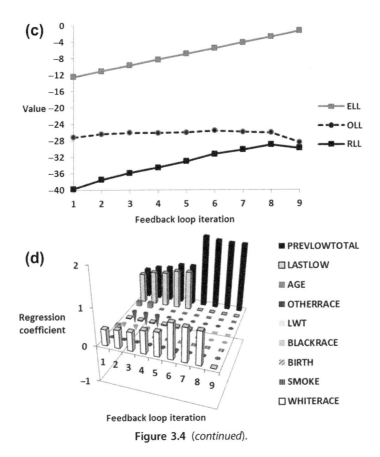

Figure 3.4 (*continued*).

categorical variables[38] and why WHITERACE, BLACKRACE and OTHERRACE are shown as distinct features. The benefits of modeling all such levels are that an effect can be estimated for any particular level of a categorical variable.[39]

In this Explicit RELR feature selection shown in Fig. 3.4(a), the *OLL* value is stable across this entire trajectory. This Explicit RELR feature selection returned only one independent variable feature which was PREVLOWTOTAL in its maximal *RLL* solution at the 9th iteration. More generally, as Explicit RELR drops features the *OLL* value may also begin to drop substantially as less accuracy is attained. In such a case, the maximal *RLL* value across all iterations will occur with more features than just the one feature in this example. Yet, the basic trajectory pattern will be the same as shown in Fig. 3.4(a) as the slowest and most stable possible trajectory to reach the maximal *RLL* in this trajectory will make it likely that the most possible features are dropped prior to reaching the maximum. Note that there was actually a local maximum in this particular trajectory at the 7th iteration. Appendix A6 presents more details on this Explicit RELR model including a presentation of key parameters like χ and β values across iterations, a demonstration of how these key parameters change at a large training sample, along with a comparison of the model's odds ratio to that reported by Hosmer and Lemeshow.

It was not possible to compute standard logistic regression models for comparison here as standard logistic regression did not converge with such highly multicollinear features like PREVLOWTOTAL and LASTLOW. But, it was possible to compare with RELR models where the least important feature is now dropped based upon χ^2 as is typically done in standard logistic regression backward selection. Figure 3.4(c) and (d) show how this feature selection proceeded with the identical training data where χ^2 is now used to measure feature importance rather than Explicit RELR's magnitude of the χ^2 derivative. In all other ways, the procedures for the model shown in Fig. 3.4(c) and (d) were identical to the Explicit RELR model shown in Fig. 3.4(a) and (b) including the identical RELR error modeling. Figure 3.4(c) and (d) show that this χ^2 measure of importance produces a different model compared with that shown in Fig. 3.4(a) and (b). This maximal *RLL* value now occurs at the 8th iteration and the corresponding model includes both WHITERACE and PREVLOWTOTAL as features. Even in this simple toy model with only nine original features, it is apparent that the regression coefficients in Fig. 3.4(d) are less stable at earlier steps like across the 5th and 6th iterations compared with Explicit RELR in Fig. 3.4(b). In fact, Explicit RELR's selection from Fig. 3.4(a) also had a significantly larger *RLL* value than the selection based upon χ^2 in Fig. 3.4(c) in an independent cross-sectional holdout sample with 60 balanced observations (-38.67 versus -42.10, $p<0.05$).[40]

So as predicted, Explicit RELR generates a larger *RLL* which reflects a more accurate model with fewer features in an independent representative sample. It should be noted that Explicit RELR may not always generate a larger within-sample *RLL* value compared with arbitrary methods to select features, especially in a very small training sample like in this example where Explicit RELR's *RLL* value was actually marginally less (-29.89 versus -28.97). However, Explicit RELR still generalized better to a new independent holdout sample than the arbitrary selection based upon χ^2. From the perspective of optimization theory, Explicit RELR is designed to generate the solution that gives a value of *RLL* that is most likely to be maximal in independent representative samples. Additionally, like all RELR methods, Explicit RELR is subject to the whims of probability. So, there also might be independent representative samples where Explicit RELR does not always give the maximal *RLL* solution, although this is also expected to be much less likely with larger training samples.

With only one independent variable feature and no intercept as in this Explicit RELR feature selection example, RELR generally gives exactly the same solution as standard logistic regression. So this is the one special case where RELR and standard logistic regression are identical in their solutions assuming that the same one feature had been manually selected in the case of standard logistic regression. When multiple features are selected, RELR still is expected to behave in a similar way to standard logistic regression in one important way. This is that RELR's regression estimates are expected to be consistent in the sense that they converge to unbiased estimates as the sample size grows at the rate of $1/\sqrt{N}$ for any model that is correctly specified, where N is the training sample size.[41] While RELR models are expected to converge to being consistent at the same rate as other maximum likelihood methods like standard logistic regression, RELR is also expected to start much closer to the very large N behavior of standard logistic regression in very small samples and higher dimensions.

Of course, a significant question that determines if consistency and corresponding lack of bias is actually observed is whether the model is correctly specified. With multiple features, RELR models are quite different from standard logistic regression in that they have much more efficient parameters with substantially lower standard errors, and are much more stable across independent samples when features are omitted or are spurious or multi-collinear. Such stability results from RELR's error model which allows much greater room for error in model specification, as error from omitting a causal feature and/or including spurious features is handled like all other error in RELR with an equal probability of being positive and negative across features. So a poorly specified Explicit RELR model still may not be that far from being "correctly specified" for those features which are in fact causal features. Namely, those causal features which are correctly included in the model still may have reasonable regression coefficient estimates. This is because the RELR error model takes into account the possibility of error due to inclusion of spurious features or omission of causal features. Still, if causal inference is required, the β parameters in Explicit RELR need to be tested in randomized controlled experiments or independent matched samples to confirm that they are relatively unbiased and unlikely to be spurious, and that is the subject of the next chapter on causal reasoning.

The *RLL* objective of Explicit RELR can be compared with other methods which are designed to introduce parsimony such as LASSO, LARS and Stepwise methods.[42,43] Unlike these other methods, this *RLL* objective is based upon an error model which differs for each individual feature selected in the model, and there is nothing comparable in these other methods to introduce parsimony. Available data do suggest that Explicit RELR may be more accurate than Stepwise models using AIC (see Fig. 3.3(a) and (b) for example). Also, Explicit RELR's feature selection often may select fewer features than Stepwise or comparable methods as suggested in Appendix A6 in reference to the Low-Birth Weight transition model of Hosmer and Lemeshow. This pattern was also reported in models that predicted first year achievement at a major urban college. Stepwise AIC selection returned almost four times the number of variables as Explicit RELR, but Explicit RELR had more accurate predictions using Brier score average squared error.[44] This same study compared Explicit RELR with other methods including LASSO, LARS, and Random Forest selection and found that RELR also did better on average squared error accuracy and was at least as parsimonious as LASSO and LARS and more parsimonious than Random Forests with variable selection. So, available evidence suggests that Explicit RELR has returned models that are simultaneously relatively parsimonious and relatively accurate compared with what are perhaps the most popular of the standard methods to introduce parsimony.

Figure 3.2(c) and (d) show the Explicit RELR feature selection in models developed from survey data on the weekend prior to the 2004 US presidential election generated by the Pew Research. This Explicit RELR model selected 9 independent variable features from a total set of more than 1000 features including two-way interactions and nonlinear effects. The very same model was achieved with a much larger set of candidate features when three-way interactions were allowed as candidates. Besides the stability which was also very clear in Implicit RELR's feature selection based upon the same data set (Fig. 3.2(a)), the most remarkable aspect of these Explicit RELR models are that there is very good face validity in the selected features. The 2004 US presidential election was all about the Iraq War and the War on Terror and those are the same features that the Explicit RELR model discovers as its most important putative causal features. The other selected features have similar validity. These features can be contrasted with the most important features from the L2-Penalized

Regression method in Fig. 3.2(b). Clearly, those L2-penalized features are very hard to interpret and involve many nonlinear and interaction effects. As shown in Fig. 3.3(a) and (b), the Explicit RELR solution is almost as accurate as Implicit RELR.

Like Implicit RELR, Explicit RELR learns to generate stable probability estimates that predict binary outcomes given input features. Unlike Implicit RELR which is only concerned with accuracy in its probability learning, the objective in Explicit RELR is to find a probability generating function that is both relatively parsimonious in terms of its selected features and relatively accurate. Explicit RELR can be interpreted to be an estimate of the most probable and most stable feature selection that meets this objective. Like Implicit RELR, the theory behind Explicit RELR is based upon stability as a most important goal in logistic regression modeling. Stability is also a very important consideration in RELR's KL updating that can be applied to longitudinal data and be influenced by its memory of its probability estimates in previous historical observations. Stable models that have stable parameters and feature selections are much more likely to replicate across independent samples and researchers. These RELR methods produce stable models not only because they take into account estimates of errors-in-variables but also because they do not have any arbitrary modeling parameters that are susceptible to human bias.

As has been discussed previously, Explicit RELR's feature selection is often parsimonious and interpretable enough to be consistent with simple putative causal hypotheses. In many cases, randomized controlled experimental methods may be available to test the putative causal hypotheses that are generated by Explicit RELR. Such experimental tests should be performed whenever possible. This is because Explicit RELR will return spurious effects without proper experimental controls. For example, Explicit RELR probably would have selected housing price as a negative risk factor for future loan defaults in the United States in 2007 without proper controls on confounding factors. Yet, in many cases in business, medicine and in many other fields, randomized controlled experimental tests are simply not possible. So other means are necessary to test any putative causal feature effects by controlling confounding factors in what may be generally called quasi-experimentation. It turns out that RELR also may be useful to generate quasi-experimental matched sample tests in causal reasoning based upon observation data, and this is explored in the next chapter.

CHAPTER 4

Causal Reasoning

"I have not been able to discover the cause of those properties of gravity from phenomena, and I frame no hypotheses. For whatever is not deduced from the phenomena, is to be called an hypothesis; and hypotheses, whether metaphysical or physical, whether of occult qualities or mechanical, have no place in experimental philosophy. In this philosophy particular propositions are inferred from the phenomena, and afterwards rendered general by induction."

Isaac Newton, The General Scholium, Principia Mathematica, 1713.[1]

Contents

In this famous passage that was appended to *Principia* in the 1713 version, Newton makes clear that he does not know the causal mechanism behind gravity and that he would prefer not to hypothesize about any such mechanism. Given the available data, he suggests that gravity by itself can be considered as the actual cause of diverse phenomena such as motions of planets and stars and seas. In essence, Newton is rejecting the more theoretical and philosophical approach to explaining nature that his rival Leibniz used which was to start with speculative hypotheses and then apply such principles to observations. Instead, Newton believed in a data-driven approach which was based upon very simple axiomatic assumptions which are "deduced from the phenomena" along with inductive observations to form very general explanatory models. Gravity may be a cause in-and-of itself or it may be essentially a proxy for an unobserved latent mechanism. Newton prefers not to speculate about either possibility, as his scientific attitude is only to stay with what he knows deductively in

Calculus of Thought
http://dx.doi.org/10.1016/B978-0-12-410407-5.00004-0

the sense of axiomatic assumptions or what he can observe inductively with data.

While Newton and Leibniz may have been on a par as mathematicians, Newton is clearly regarded to have been a much better scientist than Leibniz. In fact, Leibniz is much better known as a philosopher than a scientist. This is because the use of speculative hypotheses as a starting point to discovery does often look much more like philosophy, because it is clearly subject to similar subjective biases that are not seen in data-driven modern science.

Most causes that are discovered in science and business today are likely to be similar to gravity in that they may be proxy manifestations of deeper mechanisms. To this day, gravity is still not well understood in terms of more fundamental mechanisms. Significant theory in physics today even suggests that gravity simply could be a reflection of entropic mechanisms given by the Second Law of Thermodynamics.[2] Thus, given how difficult it has turned out to understand the mechanisms behind the causal force of gravity in the Newtonian macrophysical world that behaves as regularly as a clock, it is unlikely that detailed causal mechanisms for poorly understood human behavior outcomes like crime, poverty, sexual orientation, entrepreneurial success, or divorce will be discovered any time soon. Still, this does not mean that causes for such phenomena cannot be discovered. It is just that the causes that are discovered likely will be reflections of underlying latent mechanisms that will not be understood in any detail.

So an understanding of mechanisms will not be a likely advantage of experimental validation of putative causal effects. But, what is the advantage then? The biggest advantage is that experimental results are replicated much more often. For example, meta-studies in medicine suggest that effects that are validated through the largest randomized controlled experiments replicate most of the time, whereas studies that use standard data mining methods without experimental controls rarely replicate.[3] Hence, experimentally validated effects can be expected to give much more reliable predictions that will generalize much better to new samples of data. On the other hand, effects reported through data mining models will not generalize well especially when predictive environments change in phenomena like human behavior. So, for this reason, explanatory models where the causal inference has been tested through randomized controlled experiments should be greatly preferred.

Unfortunately, randomized controlled experiments will not be possible much of the time in human behavior or medical studies. This is not only a function of the much greater cost involved in randomized experiments, but also because many causal variables in human behavior and medical outcomes cannot be manipulated experimentally in randomized controlled studies. For example, a researcher cannot view gender, race, social class, sexual orientation, income, age, or any other person variables as variables that can be manipulated experimentally. On the other hand, quasi-experiments always will be possible with such person variables. Quasi-experiments are also very inexpensive and simply require the same general types of modeling tools as required in observation predictive modeling. Yet, when done well they can generate reliable explanatory models that allow causal inference under a critical assumption that all confounding effects have been properly controlled.

1. PROPENSITY SCORE MATCHING

Quasi-experimental methods may be most broadly defined to mean all attempts to achieve controlled comparisons as is done in randomized controlled experimental methods, but with the use of observation data. That is, quasi-experiments do not randomly assign a treatment condition to different groups because data are from naturally occurring observations. Some researchers may distinguish matching methods from other quasi-experimental methods as an argument can be made that matching methods might be closer to randomized controlled experiments in some abstract sense of degree of control over potential confounds. Matching designs are used in epidemiological research to evaluate the effect of a putative causal variable such as smoking on an outcome such as lung cancer. As an example of a matching design, the smoker and nonsmoker subgroups corresponding to the levels of the putative causal variable, smoking status, could be equated in terms of all possible covariates that could potentially cause lung cancer like gender, age, family and individual history of cancer. It would then be determined whether this selected group of smokers had a greater tendency to develop lung cancer than the matched group of nonsmokers.

While it is easy to match samples in terms of a small number of covariates like age, gender, and no previous cancer history, it is much more difficult to match in many covariates as would occur in high-dimensional data. So, multivariate propensity scores that reflect probabilities of confoundedness of covariates with outcome variables are often used instead with high-dimensional data. The most popular propensity score matched sample methods are based upon stepwise logistic regression modeling designed to predict levels of the putative causal variable in terms of all covariates that also have potentially causal effects on the outcome variable. The effect is to design exposure and control groups that are well matched in all covariates except the exposure.[4] This proceeds as follows: (1) take all covariates (gender, age, and previous cancer history) that could exert an independent causal effect on the ultimate outcome (lung cancer); (2) construct a stepwise logistic regression model which predicts the levels of the putative causal/exposure variable (smoker versus nonsmoker) based upon all these covariates; (3) construct an independent sample that matches these exposure subgroups (smoker and nonsmoker) on this propensity score estimate of smoking status; and (4) determine if these matched groups differ in terms of the outcome (lung cancer).

Widely used propensity score matching methods are essentially a two-stage regression modeling process. In the first stage, a regression model is built to get the propensity score to predict exposure groups based upon the covariates; in the second stage, a regression model is built to predict the outcome. As Imai has commented,[5] a real challenge with propensity score methodology is to get a reasonable estimate of the propensity score in the first stage, as the second stage outcome model is easily biased by even a slight misspecification of the propensity score model. A big problem is that a correctly specified propensity score model requires a high-dimensional model that traditional regression modeling has difficulty achieving. In a candid assessment of the propensity score research literature, Imai refers to the fact that balanced covariates with equal variance are the determining factors in the researcher's decision that there is a correct specification of the propensity score model as the "propensity score tautology" because the propensity score is only appropriate if it in turn balances covariates. Another problem is that model misspecification is still always possible even when this balancing occurs, so balancing is not a definitive proof of a correct specification. This is because small changes in the specification still may be associated with balanced covariates yet they also could have pro-found influence on the ultimate estimate of effects in the outcome model. This is the problem that has been reviewed in earlier chapters of this book which is that highly random changes occur in standard logistic regression stepwise regression simply due to different representative training samples or different arbitrary modeling parameters. Another related problem will occur when covariates are highly collinear with the exposure variable, in which case balance in matching may never be achieved unless the sample size is astronomically large.

Austin has provided evidence that weighting rather than matching is effective in a doubly robust procedure that weights observations by the inverse probability that they would receive the treatment in the population.[6] This requires that one knows the population parameters in order to correctly perform weighting. With weighting it is possible to achieve balanced covariates, but this necessarily gives added weight to certain observations which causes greater variance in estimates. This procedure is called *doubly robust* because it leads to unbiased results if either models that predict the probability of exposure (propensity score models) or models that predict the probability of outcome (outcome models) are specified correctly. The doubly robust procedure would avoid an incorrectly specified propensity score model, but it still requires that outcome models are correctly specified. This still may be difficult to achieve in practice.

This problem concerning correct specification of models using propensity score methods came to full light in an exchange between causal theorist Judea Pearl and his supporters versus Donald Rubin's supporters. This debate played out in the medical statistics literature and associated blogs in the late 2000s and still is an ongoing issue in the propensity score research literature. This exchange originated with a review article by Rubin on his propensity score methodology published in *Statistics and Medicine* in 2007.[7] In that article, he listed variables which his stepwise logistic regression method selected in a propensity score model that he built for litigation on relationship between smoking and lung cancer years earlier. One variable in particular attracted the attention of Pearl[8] and his followers which was whether individuals reported wearing seat belts. Pearl and his supporters argued that this was not a reasonable variable because there was no reason to suppose that wearing a seat belt causes lung cancer, and the effect of including such a variable could increase bias because of an incorrectly specified model. In a dialog that ensued on Andrew Gelman's blog, Pearl wrote, "If you know that seat belts do not CAUSE lung cancer, you should use that fact in your analysis... This is the crux of the matter."[9] In the same dialog, an anonymous blogger named Phil provided a very succinct insight into how to resolve the dilemma. Phil wrote, "The thing to do (please excuse me if this is obvious) is to look at the smoking/cancer relationship separately for seatbelt-wearers and for non-seatbelt-wearers. If it's about the same—if seatbelt-wearing smokers get lung cancer at about the same rate as non-seatbelt-wearing smokers—that would be evidence that the "smoking" coefficient is causal, or at least isn't confounded by other risky behaviors (though it still could be confounded by something other that isn't related to seatbelt-wearing)."[10]

Besides the model specification problem, another problem with Rubin's propensity score matching methods is that demographic variables like gender cannot be assessed for putative causal effects. This is because Rubin's method requires the ignorability assumption which assumes missing treatment assignment values in the data collection. This does not make sense for variables like race where assignment to white and then black cannot occur in the same person, so a race condition cannot be assumed to be missing in the same individual. This is unlike a drug or smoking status variable where assignment to either exposure or control is possible for the same individual, so one condition can be assumed to be missing for each individual in an observation study. An example of an interesting question that ideally would be evaluated with causal modeling would be whether a white woman would have had a low birth weight pregnancy if she had been nonwhite. A similar type of question is whether a woman would have made $100,000 or more had she been a man. These are essentially counterfactual/hypothetical outcomes that can never be observed when the putative causal variable is a demographic variable like race or gender.

In fact, newer propensity score methods have been developed to estimate effects of putative causal variables like demographic variables that obviously could not be exposure variables in randomized controlled experiments.[11] This has been pioneered by Frölich who has developed propensity score methods that employ semiparametric and nonparametric regression methods. Yet, Frölich's methods are still propensity score methods and still require a separate propensity score model that attempts to match groups. And, it is unclear whether covariates from high-dimensional feature sets would be truly balanced in variability across the matched groups using Frölich's methods in the sense that Rubin emphasizes is necessary for propensity score matching methods to have pseudo-equivalence with randomized controlled experiments. Yet, even in the best case scenarios reviewed by Rubin in 2007, a percentage of covariates remain imbalanced after propensity score matching. In this case, weighting is required to ensure balance which again requires that something is correctly specified—either the treatment or the outcome models. Even though a case study example exists of how Rubin's method has given almost identical results as randomized controlled experiments,[12] the extent to which this may be observed more generally is certainly open to question given that propensity score methods are highly dependent upon whether the modeler has guessed appropriate modeling specifications correctly. Still, Rubin's contribution is important in outlining the necessity of objectivity in causal confirmation methods like propensity score methods that avoid back and forth data mining methods and do so by putting the outcome variable out of sight. But the ideal would be to avoid having to construct a propensity score model in the first place.

A big advantage of avoiding propensity score models altogether would be that causal features that have correlated covariates could be discovered. This is because causal features would be expected to be highly correlated with some covariates when underlying latent causes are at work, especially in high-dimensional feature candidate sets. Yet, Rubin's propensity score method would have a difficult time discovering causal features that are proxies for such latent causes. This is because the propensity score method is designed to disallow matched groups when covariates are not matched in variability. But such imbalance in variability always happens with highly correlated covariates. So it might be best to avoid the necessity of propensity score matching in the first place,[13] and this is what reduced error logistic regression (RELR) can achieve in a matched sample design that is still not propensity score matching. That is, RELR employs a multivariate matching of a different sort which matches outcome probabilities given all other covariates while setting the putative causal effect to zero. RELR's matching design may be termed *outcome score matching*.

In fact, RELR's approach is essentially what the anonymous blogger Phil described in the above referenced exchange with Judea Pearl on Andrew Gelman's blog page in 2009. The only difference is that Phil's suggestion concerned controlling for one potentially confounding covariate, whereas RELR's causal reasoning controls all covariates simultaneously by using the probability of the outcome given all other covariates as its matching variable. The key in RELR's approach is that this holding of all potentially confounding effects constant to assess a putative causal effect can be achieved at least to an approximation without substantial bias from the inclusion of spurious features and exclusion of causal features. This is due to Implicit RELR's ultra-high-dimensional modeling capabilities and its error modeling which remove biasing effects of multicollinearity and omitted features.

2. RELR'S OUTCOME SCORE MATCHING

Instead of a two-stage process that first develops a propensity score logistic regression model to predict exposure variable subgroups and then develops matched samples to determine exposure variable effects on an outcome variable in second regression model, one could theoretically achieve the same end with a regression model designed to predict the outcome variable. That is, it would be possible to construct a regression model to predict the outcome (e.g. lung cancer status) that adjusts or controls a putative causal variable (smoking status) based upon all potential confounding covariates. In fact, in some of his earlier work on matching methods, Rubin evaluated such regression covariate adjustment through simulations. These simulations suggested that regression covariate adjustment is only effective at removing bias when the exposure and control groups have roughly the same variance in their covariates. These simulations also suggested that multicollinearity and omitted nonlinear effects would likely have a pathological biasing effect on estimations of causal effects. Because of such pathological biasing, Rubin suggested that matched samples with possible corresponding regression covariate adjustment would be preferable to simple regression covariate adjustment based upon random, unmatched samples.[14] Rubin went on to refine his propensity score matching methodology which requires correct specification of two models—one to predict the exposure groups and one to predict the outcome.

Explicit RELR might overcome some of the bias issues that concerned Rubin with regression and covariate adjustment based upon random samples that are not matched. This is because Explicit RELR overcomes the pathological issues that Rubin observed related to multicollinearity and nonlinear effects. However, Explicit RELR also can be more subject to error effects than Implicit RELR, especially when Explicit RELR selects just a few features in which case the assumption that positive and negative error completely cancel without bias across features may be less valid. For this reason, it is very important that extraneous error due to biasing effects be controlled in Explicit RELR as much as possible if it is to be used to select putative causal features.

In fact, there is one special case where Explicit RELR will avoid bias due to how variance is computed in the t values of the error model and due to the intercept term which is not modeled with the error model. This is the special case when the training sample is not random but instead stratified into target and nontarget groups that are perfectly balanced so that the numbers of target and nontarget values in the outcome variable are equal and where the intercept is also excluded. The importance of this special case is reviewed in Appendix A7, as the model will avoid bias due to how variance is estimated in the computation of t values and by the inclusion of an intercept in perfectly balanced samples.

Yet, another problem with viewing effects returned by Explicit RELR as unbiased is that there still could be unequal variances in covariates across a putative causal factor like smoking and nonsmoking groups. As noted above, Rubin showed that regression covariate adjustment is only effective when there is equal variance in covariates across a causal factor. So any interpretation that an Explicit RELR regression coefficient reflects an unbiased causal effect while holding all other effects constant should be viewed as a question or more appropriately a hypothesis that may be falsified with further data.

So, there needs to be a quasi-experiment that tests whether a given selected Explicit RELR standardized regression coefficient reflects a causal effect while holding all potentially confounding effects constant. This is done for each effect that is selected in Explicit RELR based upon a perfectly balanced sample and a zero intercept. That is, each selected feature can be viewed as a separate putative causal effect which leads to a separate hypothesis-testing process while holding all other factors constant. The fine details of RELR's outcome score matching method are listed in Appendix A8, but a higher level overview is summarized here.

In the first step, an Implicit RELR Offset model is developed which includes the hypothesized causal effect obtained from Explicit RELR as an offset feature. For each observation, this offset feature value is the product of the Explicit RELR regression coefficient and its standardized feature value. The offset variables here will have a constant coefficient of one for each data observation and a coefficient of zero for each pseudo-observation. This is because the offset effect is a fixed value obtained from the previous Explicit RELR model, so it is treated as a fixed value here where the RELR error model has no effect on it. Just like RELR's error models when there is no such offset, it is expected that all error across all included covariates in the model will have an equal probability of being positive and negative without bias. This Implicit RELR Offset model should also now include all relevant covariates which should be all features that remained after feature reduction in the original Explicit RELR model from which putative causal features are derived, or all features included in that Explicit RELR model in a low-dimensional candidate set that has no feature reduction. This Implicit

RELR Offset model is ideally a very high-dimensional model which controls for all other features while achieving an estimate of the putative causal effect.

Once this Implicit RELR Offset model is built, the levels of the putative causal variable (e.g. smoker cases versus nonsmoker controls) are matched in terms of the probability that all other covariates contribute to the outcome. The probability score that is used for matching can be called the *covariate outcome probability*; it is the probability of the outcome variable given all features other than the putative causal effect. This matching is designed to yield two groups: one is a case group like smokers and the other a control group like nonsmokers, which have equal variance in such covariate outcome probabilities. In the case of a putative causal feature that is a numeric variable like income or age, the expected value that is used as the cutoff may be the average.[15] Values above and below such a cutoff can be used to form the case and control subgroups. Any interaction or nonlinear features that had been selected in the Explicit RELR model also should be evaluated separately as separate features from any linear or main effect. In this matching of the case and control subgroups that ensues in terms of the covariate outcome probabilities, the convention is followed that the standard error of the difference measure should be below 0.25 in this covariate matching variable for each group.[16] This matching of the standard error of the difference serves to ensure that the covariate outcome probabilities have roughly equal variance in the case and control matched groups.

There are a number of approaches to matching probability distributions,[17] and all will likely give similar results when the goal is to find a set of matched pairs with minimal differences. Yet, the approach that is used in RELR's causal reasoning is the Topsøe distance.[18] The Topsøe distance is a symmetric version of the Kullback–Leibler (KL) divergence that was introduced in the last chapter.[19] The Topsøe distance $T(\mathbf{p},\mathbf{q})$ can be defined based upon the KL divergence given in Eqn (3.2) as follows:

$$T(\mathbf{p}, \mathbf{q}) = D(\mathbf{p}, \mathbf{m}) + D(\mathbf{q}, \mathbf{m}) \qquad (4.1)$$

where each element within the probability distribution $\mathbf{m}$ is given by $\mathbf{m} = (\mathbf{p} + \mathbf{q})/2$ and where $D(\mathbf{p},\mathbf{m})$ and $D(\mathbf{q},\mathbf{m})$ are the KL divergence values for the case and control covariate outcome probabilities corresponding to above and below the expected value of the putative causal feature that is tested relative to their average value $\mathbf{m}$. It is arbitrary whether the case or control is called $\mathbf{p}$ or $\mathbf{q}$ in this case, as the exact same answer will be found either way due to the symmetric nature of the Topsøe distance. Each case/control pair is matched individually in a 1:1 matched pairs design.

The Topsøe distance is used because it is more understandable in terms of information theory than all other potential measures. The Topsøe distance can be thought of as a comparator that varies between 0 and 1.38 depending upon the degree of mismatch that exists between two outcome probability distributions. This Topsøe matching signal is zero when there is a perfect match, whereas it is 1.38 when there is a perfect mismatch. In the next chapter, the neural analogs of RELR will begin to be introduced, but for now it should be noted that the close relative to the Topsøe distance known as the Jensen–Shannon divergence which is simply half the Topsøe distance defined here has received significant attention in neural coding research. One application for example has been to cluster populations of neural signals into subgroups based upon whether they represent matching or nonmatching signal patterns in retinal ganglion cells of the salamander, which is very similar to the present application.[21]

The KL divergence is asymmetric and typically represents the sequential updating of a temporally prior distribution q into a temporally posterior distribution p. The KL divergence has meaning in terms of representing the information that is gained through such sequential update learning across time. In contrast, the Topsøe distance is symmetric and represents the convergence of two simultaneous distributions p and q into an average distribution m. The Topsøe distance has meaning in terms of the degree of mismatch in the original p and q distributions. The minimum KL divergence and minimum Topsøe distance generally define two fundamental objectives corresponding to whether outcome observations that are the basis of distinct probability distribution groupings used in RELR are measured sequentially as in the case of minimal KL divergence or simultaneously as in the case of the minimal Topsøe distance.

Once a quasi-experiment is all set up to yield two subgroups that match based upon Topsøe distance, a simple hypothesis test is then performed to determine if there is a significant difference between these two matched groups in terms of the outcome (e.g. lung cancer versus non–lung cancer). RELR's preferred approach to test binary outcome differences between matched groups is conditional logistic regression; McNemar's test of correlated proportions is also often used for similar applications and does give similar results to conditional logistic regression.[22] It is important that the test is one that is appropriate for dependent groups, as is the case with both conditional logistic regression and McNemar's test. Conditional logistic regression is used in RELR because it is identical to RELR in this matched sample testing as shown below. If the p value that results is below a predetermined significance level which is <0.05 in a one-tailed test often in biomedical applications, consider this evidence in support of the causal hypothesis. Otherwise, the conclusion would be that the causal nature of the relationship is not supported by the test.

Just as in Rubin's objective propensity score matching methodology, all research design decisions should be made in advance of viewing the outcome data used in the test. This includes, for example, the significance level for the conditional logistic regression, the candidate features to be included in the original Explicit RELR model, the sample used for the Explicit RELR and Implicit RELR Offset models, the features to be tested as putative causal features along with their cutoff levels to form case and control groups, the independent sample used for the conditional logistic regression test, and the acceptable level of the standard error of the difference that is used to match groups for this conditional logistic regression test. While pilot data may be helpful without longitudinal data, the independent data used in this conditional logistic regression will need to be future data in a longitudinal sense for a more definitive test with causal factors that can be present or absent over time. This is because it is extremely important to demonstrate that the putative causal factor was present prior to its effect to be consistent with an important aspect of causality which is that real causes must precede outcomes that they cause.[23]

The only nuance here concerns conditional logistic regression. Readers may wish to consult a standard logistic regression text to get an overview of conditional logistic regression.[24] What can be said here is that conditional logistic regression is unlike standard logistic regression in that the groups that are compared are not independent groups, but instead are dependent groups as would happen in a case–control matched sample grouping or a before-and-after experiment. Because of this, it has rather different properties than standard logistic regression. However, in a special case of only two matched groups known as 1:1 matching, conditional logistic regression can be set up just like standard logistic regression where a new dependent variable is now everywhere equal to one and where there is not an intercept. In this special case, the difference between the original outcome events (e.g. lung cancer or no lung cancer) in each matched pair now serves as the independent variable in this model when it is viewed in analogy with standard logistic regression. In the special case of how conditional logistic regression is applied in RELR's causal reasoning, there is only one independent variable feature which is this difference between the original outcome events in the matched pairs. And RELR will give exactly the same solution as conditional logistic regression and standard logistic regression in this case which also includes no intercept. As reviewed in Hosmer and Lemeshow, the conditional logistic regression coefficient in such a very special design of 1:1 matched pairs is simply the log of the ratio of the discordant pairs.[25]

3. AN EXAMPLE OF RELR'S CAUSAL REASONING

The Explicit RELR model that serves to generate the putative causal hypothesis in this example is the model shown in Fig. 3.4(b) and is described in the previous chapter. This model was based upon the same cross-section of data from the Low Birth Weight dataset that was used to build the Explicit RELR model of Fig. 3.4(b). That Explicit RELR model had selected PREVLOWTOTAL as the only putative causal feature with roughly an odds ratio of 10 for a low birth weight pregnancy given one previous low birth weight pregnancy compared to none at all. However, there is reason to question this odds ratio given that maximum likelihood logistic regression is known to return regression coefficients that are biased to be high in magnitude in small observation training samples. So, it would not be surprising to find a less biased estimate through the causal matched sample modeling.

Figure 4.1 shows the Implicit RELR Offset model that was developed based upon the product of the standardized PREVLOWTOTAL feature and its regression coefficient as the offset variable, and the same original covariates in the Explicit RELR model shown in Fig. 3.4(b). The outcome variable for this model was the same Low Birth Weight variable used in Fig. 3.4(b), and the training sample used to construct this model is also exactly the same training sample used for the model in Fig. 3.4(b). Note that the offset variable actually received a weight of one for the observations and zero for the pseudo-observations in this offset regression, as the regression coefficient of 1.84 shown in Fig. 4.1 for PREVLOWTOTAL is from the original Explicit RELR model. All covariates had regression coefficients that were at least 1.49 times larger in magnitude in the model given in Fig. 3.4(b), so this suggests that there is a reduction of bias in these covariate weights in this Implicit RELR Offset model shown in Fig. 4.1. This is especially true of the LASTLOW covariate which is roughly 0.41 in this model, whereas it was roughly 0.78 in the first iteration of the model shown in Fig. 3.4(b) which included these identical covariates.

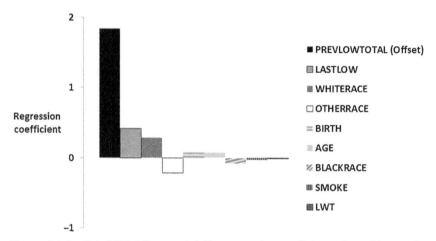

Figure 4.1 Implicit RELR Offset model. The regression coefficient of roughly 1.84 for the PREVLOWTOTAL from the original Explicit RELR is also shown, although PREVLOWTOTAL was treated as the offset in this particular model.

The bias is also less in the PREVLOWTOTAL effect when evaluated based upon a matched sample that uses the covariate outcome probabilities as its matching variable. This is shown in Table 4.1 in the conditional logistic regression which tested the PREVLOWTOTAL effect in the matched sample. This sample that is evaluated in this model in Table 4.1 is made up of observations that are completely independent from the training sample used in the Implicit RELR Offset model. That is, these not only were separate patients but also were evaluated at a separate and later time study from the sample used to construct the Explicit RELR and Implicit RELR Offset models. This sample was the last cross-section in the Low Birth Weight dataset, and had 133 total observations (40 Low Weight Births and 90 No Low Weight Births). While this model returned a regression coefficient of 1.95, this is different from Explicit and Implicit RELR models in that this is not a standardized regression coefficient, as the case–control outcome differences here which are effectively the independent variable are not standardized variables. In fact, the odds ratio that a woman who had a previous low weight birth will have a low weight birth compared to a woman with no previous low birth weight history is now only 7, whereas it was 10 for having had one previous low birth weight pregnancies and 516 for having had two previous low birth weight pregnancies relative to having had none in the Explicit RELR model used

Table 4.1 Key Parameters in Low Birth Weight Case–Control Conditional Logistic Regression. Cases had PREVLOWTOTAL > 0; Controls had PREVLOWTOTAL ≤ 0. Covariate Outcome Probabilities were based upon the Implicit RELR Offset model but were the results of scoring the last cross-sectional sample using that model as the probability of the outcome (Low Birth Weight) given all other covariates shown in Fig. 4.1 while excluding the offset PREVLOWTOTAL. Topsøe Distance and Squared Difference are each shown for each matched pair to demonstrate that these measures give very similar values for small differences as in this application. The regression coefficient β was computed based upon the natural log of the ratio of discordant Case–Control pairs in the outcome, which was $1.95 = \ln(7)$. The probability p in the far-right column was the logistic regression outcome and is $p = \exp(1.95x)/(1 + \exp(1.95x))$ for each matched pair, where x is simply the difference between the Case and Control Outcome that is shown for each pair. Note that because matches are ordered in terms of distance of matches, the last matches are the most distant as is apparent in this list.

Case covariate outcome probability	Control covariate outcome probability	Topsøe distance	Squared difference	Case outcome	Control outcome	p ($\beta = 1.95$)
0.401	0.401	8.63E-09	8.29E-09	1	0	0.875
0.459	0.459	2.45E-08	2.43E-08	1	1	0.5
0.429	0.429	3.35E-08	3.28E-08	1	0	0.875
0.401	0.401	1.24E-07	1.19E-07	1	1	0.5
0.414	0.415	5.19E-07	5.04E-07	1	0	0.875
0.403	0.403	6.03E-07	5.81E-07	1	0	0.875
0.450	0.450	6.10E-07	6.04E-07	0	0	0.5
0.426	0.425	6.35E-07	6.21E-07	1	1	0.5
0.479	0.478	1.21E-06	1.21E-06	1	0	0.875
0.422	0.420	3.54E-06	3.45E-06	0	0	0.5
0.465	0.463	3.54E-06	3.52E-06	0	1	0.125
0.414	0.417	6.69E-06	6.50E-06	1	1	0.5
0.467	0.463	2.03E-05	2.02E-05	1	1	0.5
0.423	0.419	2.26E-05	2.20E-05	0	0	0.5
0.445	0.440	2.29E-05	2.26E-05	1	0	0.875
0.510	0.498	1.62E-04	1.62E-04	0	0	0.5
0.592	0.548	1.98E-03	1.94E-03	1	0	0.875
0.635	0.481	2.42E-02	2.38E-02	0	0	0.5

to generate the causal hypothesis from Fig. 3.4(b). Thus, relative to the Explicit RELR model, there is a trend for a lower bias in the present odds ratio estimate. A chi-square test can be set up by calculating the log likelihood of each matched pair observation both with $\beta = 1.95$ and with

$\beta = 0$ and then summing across observations, and then computing minus two times the difference in log likelihood values between these two conditions. This test showed that this effect was significant ($\chi^2 = 5.06$, df $= 1$, $p < 0.05$).

One caveat is that the case sample here had only 3 out 18 (16%) total cases that had two previous low birth weight pregnancies and none that had more than two whereas the total set of 61 cases from which this sample was constructed had 18 out of 61 total cases (29.5%) with two previous low birth weight pregnancies and 3 out of 61 total cases (4.9%) with three. The rest in both samples only had one previous low birth weight pregnancy. With a larger sample it would be possible to select the number of cases better so that they match better the percentage in the larger sample. In this way, better support and generalization would be observed as the sample of cases used in the matched sample would be a better match to the general population. A decision about whether or not to do this should be made prior to observing the case–control matched sample results like all other design decisions.

Obviously with very small samples like in the present example, statistically significant results need to be interpreted with caution. One nuance of this matched sample conditional logistic regression test is that the regression coefficient calculation based upon the ratio of discordant pairs only can be computed when there are in fact both types of discordant outcome pairs (case $= 1$ and control $= 0$) and (case $= 0$ and control $= 1$). That was the situation here, but there was only one discordant outcome with case $= 0$ and control $= 1$ in the present example. A much smaller effect would have been obtained with two such discordant outcomes, which would not have been significant.

The standardized error of the difference in the case and control covariate outcome probabilities was roughly 0.23 in the data in Table 4.1. The matched pairing is shown in the same order that it was entered based upon the minimal Topsøe distance between the case and control covariate outcome probabilities. The next such matched pairing put the standard error of the difference over the limit of 0.25. Very similar results would have been obtained if squared difference had been used. In fact, any symmetric metric that accurately measures small differences between probability distributions should give very similar matched groups in this application. Unlike Rubin's propensity score methodology, there is not an attempt to ensure that all covariates are equated in variance across the case and control subgroups. In fact, covariates that are highly correlated to

PREVLOWTOTAL such as LASTLOW in the present example will have substantial differences in variability across the two matched groups. And such differences in variability would be expected if LASTLOW's effect on the outcome reflects the same underlying latent causal factor as PREVLOWTOTAL's effect (see below). Yet, LASTLOW is not used to match these two groups as instead the covariate outcome probability is used, and this covariate outcome probability does have roughly equal variance in these two matched groups.

The Implicit RELR Offset model that is required to produce the co-variate outcome probabilities in RELR's outcome score matching would not be generally possible with standard logistic regression. This is because standard logistic regression has substantial bias and convergence problems with high-dimensional multicollinear features. Even when a solution is found in standard logistic regression with such data, it usually gives very unstable regression coefficients that are biased to be high in magnitude along with probabilities that randomly vary as a function of the training sample and as a function of other covariates that are entered into the model. In contrast, Implicit RELR generates solutions that are very stable in terms of regression weights and probability scores across independent samples and as a function of other covariates entered into the model. This stability typically increases as more covariates are put into a model. Any tendency for magnitudes of covariate regression coefficients to be biased high also decreases with higher dimension covariates as the magnitude of the RELR regression coefficients is substantially less with high-dimensional versus low-dimensional features selected as seen in Fig. 3.2(a) and (b) for example. While individual covariate regression coefficient magnitudes are obviously biased to be low in high-dimensional RELR models, these individual coefficients are not interpreted and instead only the combined covariate outcome probability estimates are used which are not expected to depend significantly on the inclusion or exclusion of individual covariate effects.

Thus, the probability estimates in the Implicit RELR Offset model are unlikely to be strongly biased by the inclusion or omission of a controversial covariate such as seat belt wearing in a model designed to predict the effect of smoking status on lung cancer. In any case, great care always should be taken to ensure that the original Explicit RELR predictive model that is used to generate causal hypotheses includes as many candidate features prior to feature reduction as possible as ideally reflected by the opinion of several experts, along with the opinion of the modelers who may not be subject-matter experts. This should also include interaction and nonlinear features that are built from such simpler features. But in cases where there is controversy between subject matter experts or modelers over whether a simple feature such as seat belt wearing should be included, the feature always should be included because it may be a causal feature that would be selected in the Explicit RELR model and supported in the causal testing.

While the PREVLOWTOTAL feature discovered here is a likely causal risk factor for Low Birth Weight pregnancies, it does not actually represent a manifest causal feature. Instead, this type of variable reflects an unknown latent cause. That is to say that PREVLOWTOTAL is simply a proxy measure for an underlying mechanism that actually causes the outcome here of low birth weight pregnancies. A similar argument can be made that seat belt wearing reflects a latent causal variable such as a personality trait associated with lower risk taking. For example, people who wear seat belts may be less likely to engage in certain behaviors such as working at a chemical plant with toxic fumes that also could cause lung cancer. Thus, wearing seat belts could be associated with a decreased probability that a smoker will get lung cancer in such a causal scenario.

The point is that nobody usually knows in advance whether a given feature may or may not reflect an underlying cause at least in outcomes related to health or human behavior, so there will be substantial disagreement over likely causal features. For this reason, it would be unwise to rule out certain candidate causal factors just because they are deemed controversial. One purpose of research should be to discover new causes that may never have been considered by most experts. Even if a controversial feature like seat belt wearing is ultimately not selected in the Explicit RELR model, the inclusion of such an unimportant candidate feature will have a very minimal effect on the high-dimensional Implicit RELR Offset model that is used to generate the covariate outcome probabilities in the matching.

RELR is also much less sensitive to the omission of a causal feature than standard logistic regression, but the cost of omitting of causal feature is greater in RELR than the cost of including a spurious feature. If the PREVLOWTOTAL feature had been omitted in the low birth weight model, its closely related correlate LASTLOW (i.e. a low birth weight in the last pregnancy) would have been selected instead in Explicit RELR. But, this was a smaller effect and would not have been as significant in the causal matching test. Also, simply knowing that a low birth weight in the last pregnancy is associated with a low birth weight in the next pregnancy is less informative than knowing that the total number of previous low birth weight pregnancies (i.e. PREVLOWTOTAL) is the causal risk factor. So, by starting with a very high-dimensional set of potentially important features, there is a much greater likelihood of arriving at causal factors that are stronger effects and easier to interpret because they more directly reflect underlying latent causal mechanisms.

The discovery of a causal factor that is a proxy for an underlying mechanism is actually very similar to what occurred in the discovery of gravity, as Newton's discovery that gravity is the cause of planetary motion only begs the next question. What causes gravity? In the present situation where the prior history of low birth weight pregnancies can be viewed as the major causal risk for future low birth weight pregnancies, studies obviously need to be done that uncover causal factors for initial low birth rate pregnancies where the prior history can be controlled. The Hosmer and Lemeshow Low Birth Weight dataset are simulated data that are designed to embody those effects that are found in real-world data. However, because they are still only simulated data, nothing can be read into the causal inference here related to the PREVLOWTOTAL effect that has any bearing on the real world.

RELR's outcome score matching is similar to the kind of causal reasoning that led us to the discovery of the long preclinical process in Alzheimer's disease. Instead of matching groups based upon average values of covariates as in our studies, RELR's outcome score matching matches pairs of cases and controls. For this reason, RELR's outcome score matching allows much greater quasi-experimental control. RELR's outcome score matching is designed to be similar to the way that scientists reason about causes, as the important idea always of a causal factor is that it has an effect on the outcome while holding all other factors constant. Of course, the success of RELR's quasi-experimental outcome score matching will depend on the extent to which all other factors were indeed held constant which will usually mean that a high-dimensional candidate set of features is used. But this is not different from randomized controlled experiments which are also susceptible to bias when all other factors are not held constant. Like randomized controlled experiments, RELR's outcome score matching is completely objective and mechanical and can be easily replicated by independent scientists with independent data. The fact that RELR's outcome score matching removes human subjective cognitive bias and endless debate about whether or not a controversial covariate like seat belt wearing should be included might be its greatest virtue.

4. COMPARISON TO OTHER BAYESIAN AND CAUSAL METHODS

RELR is essentially standard maximum likelihood logistic regression with appropriate error modeling through its pseudo-observations and symmetrical error probability constraints. Because of this relationship with standard maximum likelihood logistic regression, RELR models have the advantages of other parametric regression models which most notably are the transparent and understandable parameters. RELR is also based upon the objective Bayesian principle that Jaynes popularized that the best solution is always the most likely inference given the measurements and the uncertainty. This RELR solution happens to be the same solution that would be gotten using Fisher's maximum likelihood principle as applied to logistic regression given appropriate error modeling. This essentially removes the distinction between Bayesian and Frequentist views in RELR, as RELR embodies both views. Yet, RELR does not have the disadvantages of standard logistic regression and its many variants along with Bayesian methods that are often used in causal modeling.

4.1. Comparison to Naïve Bayes

The Gaussian Naïve Bayes classifier, or what is usually just called Naïve Bayes, is a wonderfully simple approach that often returns very accurate and stable models with very small sample sizes.[26] The reason that Naïve Bayes often works so well is that it simplifies predictive modeling problems to avoid the curse of dimensionality. The basic assumption in Naïve Bayes is one of conditional independence between all independent variable features. Conditional independence ensures that how one feature affects an outcome in no way interacts with how another variable affects the same outcome. Thus, if we know that warm weather improves our sales process, then this weather effect is conditionally independent from the age of our customers if this weather effect does not increase or decrease if the customer is younger or older. While an assumption of conditional independence effectively excludes the possibility of interaction effects, often interaction effects are entirely absent or extremely small, so they will not have an appreciable effect on predictive ability.

However, there are times when large interaction effects will be present. For example, it may be that older customers actually are more likely to buy our product on warm days, whereas younger customers are more likely to buy on cold days. Naïve Bayes will be completely blind to such an inter-action effect. Beyond not seeing interactions, Naïve Bayes forces a rigid structure to the regression coefficients that force them to be proportional to Student's *t* values that measure the effect of the independent variable on the binary dependent variable outcome. This rigid proportionality does not allow flexibility to see regression coefficients that are not tied to this rigid proportionality. Naïve Bayes can be more accurate than standard logistic regression with very small samples, but all evidence suggests that standard logistic regression will outperform Naïve Bayes with large samples even when no interaction effects are present. The reason is that standard logistic regression is able to adjust its regression coefficients in ways that are not rigidly proportional to univariate *t* value effects.[27]

Implicit RELR models have some similarity to Naïve Bayes in that their regression coefficients can be proportional to *t* values at very small samples with a large number of independent variables. Also, Explicit RELR models that select a large number of features in relatively small sample data will also have regression coefficients that may be proportional to *t* values. However, this rigid proportionality breaks down in Implicit and Explicit RELR models with larger samples and with smaller numbers of selected features. Also, all RELR models can include interaction effects, and interaction effects often can improve the accuracy of RELR models. So like standard logistic regression, RELR is not subject to the rigid conditional independence assumptions of Naïve Bayes. This is an advantage.

4.2. Comparison to Hierarchical Bayesian methods

Hierarchical Bayes methods have been introduced in many science and business applications in the past decade. In common implementations, a multivariate normal distribution is generated for the prior probability used in Bayes' Rule. The parameters for this construction are ideally based upon empirical data. Hierarchical Bayes methods typically borrow from obser-vations at a larger aggregate level to impose greater stability on a lower level. An exemplary situation described in a standard Bayesian textbook is how survival outcomes in past experiments involving a clinical intervention in rodents may be used to form the higher aggregate level prior beta distri-bution that imposes prior probabilities on the lower level involving a new experiment.[28] In a marketing science application, the higher level may be a regional or national level of retail stores, whereas the lower level may be individual stores. Often there are more parameters than data points in

Hierarchical Bayes models. Any such Hierarchical Bayesian model also can be modeled nonhierarchically. Yet, traditional nonhierarchical constructions such as standard logistic regression tend to overfit small samples to some extent compared to Hierarchical Bayes models.[29] However, Hierarchical Bayes models have their own error problems when even more than a few irrelevant multicollinear features are allowed as candidate variables in a model.[30,31] Because of these problems, Hierarchical Bayes is not an appropriate method in data mining applications with many potentially irrelevant features.

Hierarchical Bayes modeling is much more appropriate in randomized and controlled experimental settings where the independent variables can be forced to be uncorrelated to avoid multicollinearity issues. Hierarchical Bayes models are widely used in discrete choice experiments in marketing science applications. Such choice modeling can help to determine the likelihood that consumers may purchase a product when it contains various combinations of attributes, and causal interpretations are possible when data are collected in randomized controlled experiments. The mixed logit model is one such implementation and has been discussed in Chapter 2 as an example of a random effects model that is applied to correlated observations.

At an aggregate level, Hierarchical Bayes discrete choice models also may have an advantage over standard multinomial logistic regression in that they are less affected by violations of the Independence from Irrelevant Alternatives (IIA) assumption. The IIA assumption is best seen in the Red-Bus/Blue-Bus example. Suppose that people are asked whether they would prefer to take a Red Bus, a Blue Bus or a Train to commute to work and 25% prefer the Red Bus, 25% prefer the Blue Bus and 50% say they would prefer to take the Train. Next suppose that we remove the Blue Bus as an alternative in the real world. Assuming that there is no preference difference between Red and Blue Bus, standard logistic regression model will incorrectly extrapolate to this new choice scenario with a prediction that one-third will choose the Red Bus and two-third will prefer the Train. However, the actual empirical results would be that 50% would prefer the Train, whereas 50% would opt for the Red Bus. Obviously, this is an example of a poorly specified design, where the experimental procedure does not have external validity in terms of how the model is applied. In many cases, a better specified design will handle these issues. For example, the alternative could be a binary decision whether to purchase a ticket or not, and the independent variables in this decision could be the model of transportation (bus or train) and the color of the mode (blue or red) in a balanced experimental design.

The IIA restriction assumes that the removal of alternatives will not affect the relative preference proportions of existing alternatives. Otherwise, in cases where alternatives are removed, the model will not extrapolate very accurately to the new situation. Outcomes that result from multiple alternatives where we must decide only to buy one product or vote for one candidate can be strongly affected by inclusion of correlated choices. A political election where two conservative candidates run against a liberal candidate may have a different winner when only one conservative candidate is in the race against the liberal candidate.

RELR only allows binary and ordered logistic regression models to be developed, both of which do not have IIA restrictions. Separate binary models may be constructed in a multinomial choice paradigm, such as a separate model for each of two conservative candidates versus the one liberal candidate. But, the limitations imposed by IIA restrictions still have an effect on RELR modeling even when binary models are constructed from multinomial choice sets. This is because the models may not generalize to choice paradigms that include a reduced set of alternatives. One way around this is to nest the binary models, so that a model is built first for a choice between alternative 1 and alternative 2 or 3. If the chosen alternative is two or three, then a separate model is built that applies to this choice. When done judiciously, a nested model may generalize to choice paradigms that include a different set of alternatives. For example, a nested model that groups two conservative candidates together in an election poll against one liberal candidate may generalize to a situation where only one conservative candidate is in the race.

The RELR feature selection that maximally discriminates a given binary category from the reference category would be same without regard to the other multinomial categories that are represented in the model. This is an advantage as feature selection in multinomial logistic regression is arbitrarily determined by all other multinomial choices and would likely change when a new multinomial choice is entered as a possibility. RELR models also allow hierarchical variables including individual level and more aggregate level variables to be modeled simultaneously as separate features. For example, a retail model may include all regions, states, cities, and individual stores as dummy coded features, but Explicit RELR would select the small set of relatively nonredundant features that best predict the outcome at the individual person level. Like Hierarchical Bayes methods such as mixed logit, RELR will avoid problems related to correlated observations. In RELR, this results from dummy coded features that avoid multicollinearity error in the case of cross-sectional samples and through minimum KL divergence online learning in the case of longitudinal data.

So RELR would be appropriate in all modeling situations where Hierarchical Bayes methods are currently deployed although care is required to avoid IIA generalization problems. RELR models will only generalize to those representative choice sets which were used in training. If real world data may have a variable set of choices, then representative training samples need to be designed for such a possibility. But given that random effects designs like Hierarchical Bayes require assumptions only observed in experiments, RELR should be much more useful with observation data. This is not only because of RELR's ability to yield causal hypotheses based upon data mining of observation data. But, this is also because RELR gives causal testing in observation data through outcome score matching and it allows person variables like gender, age, and income to be tested as putative causal features.

4.3. Comparison to Bayesian Networks

In Bayesian Networks, conditional dependencies in the form of interactions can be modeled. But it is much simpler from a computational perspective if conditional independence can be assumed like in the form of Bayesian Networks called the noisy-OR Bayesian Network. In cases with no missing data when conditional independence is assumed between all attributes, noisy-OR Bayesian Networks can be designed to return exactly the same class labels for the same observations as Naïve Bayes given the same binary attributes.[32] However, the direction of causality in noisy-OR Bayesian Networks is from attributes to classification outcomes, whereas the opposite direction is invoked when a Naïve Bayes model is graphed as a Bayesian Network.

The noisy-OR gate is based upon the logical OR operations that give a true result when any one condition in a set of conditions is met. This gate is noisy because it gives a result that is not completely true given any one condition, but instead is only likely to be true and is expressed as a probability. As an example, assume that the probability that a person will NOT be diagnosed with probable Alzheimer's disease if they are over age 80 is 0.75. Also, assume that the probability that a person will NOT be diagnosed with probable Alzheimer's disease is 0.8 if they have had a previous head injury. In the noisy-OR model, we assume that these effects act independently like two independent coin flips, so the probability of the outcome of NOT being diagnosed with probable Alzheimer's disease is simply the product of these two conditionally independent events or $(0.75)(0.8) = 0.6$. This is like having two pipes that bring in each causal factor in an erratic way, but each factor by itself can cause the outcome. So, it is a logical OR function that is noisy. There are many other types of Bayesian Network models which make different assumptions than these noisy-OR gate assumptions.[33] As one example, interaction effects also can be modeled in some specific scenarios by introducing another probabilistic variable to account for the interaction between variables in what is known as a leaky noisy-OR model. This interaction effect may be assumed to abide by locally conditional independent probability rules, so it also may be approximated as a similar multiplicative effect when these assumptions are realistic.[34]

A major interest in Bayesian Networks is to discover causal factors with observational data when all modeling assumptions, such as conditional independence, are met. The most prominent architect of the view that well-structured Bayesian Networks can produce causal insights is Judea Pearl through his book entitled *Causality: Models, Reasoning and Inference*. The view of causality that is outlined in his book is a classical view of determinism where causes are not viewed as probabilistic, but instead are seen as forces in much the same way that macro forces in physics act on objects. The causes modeled by Bayesian Networks in Pearl's book are feed forward unidirectional causes where no feedback is involved, so they can be graphed as directed acyclical graphs. In the case of unobserved causal variables, Pearl introduces latent variables built from measured features.

Pearl's notion of causal modeling through noisy-OR Bayesian Networks is really a deterministic view of causality. This view is nicely encapsulated in the noisy-OR logic, which through the logical OR mechanism would be as determined as a clock if not for the noisy uncertainty in its response. Yet, one often cannot come to a reasonable causal model through observational data alone in complex Bayesian Networks, especially as the number of nodes gets very large. This is because the optimal or most minimal conditionally independent network structure is NP-hard to learn.[35] An NP-hard problem is a problem that cannot be solved in practical time as the problem size gets large with today's computers. Because of this NP-hard status, expert knowledge is often employed to impose constraints that can lead to reasonable leaky noisy-OR Bayesian Networks, but hand crafting is also more difficult with large problems. Thus, assumptions may need to be imposed on the network structure that really cannot be verified by experts or prior empirical data in a large variable problem. Beyond the rather significant problems in learning Bayesian network structures in problems with many potential variables, other problems relate to confounding variables and model identification.

The issue of confounding variables is best seen in Simpson's paradox. This paradox was first mentioned by Pearson in 1899,[36] but it was the main theme of a paper by Simpson many years later from whom it is named.[37] This paradox relates to how trends that are observed across group means may reverse at subgroup levels. A classic demonstration of Simpson's Paradox is seen in success rates from a clinical study[38] related to two different kidney stone treatments: open procedures and percutaneous nephrolithotomy. Open procedures are slightly more effective in both the small and large stone subgroups, but when these subgroups are combined, open procedures are less effective on average. What is happening is that the size of the kidney stone is a confounding variable. Large stones are much more difficult to treat and are overly represented in the open procedures treatments. Thus, on average open procedures appear less effective when both small and large stone groups are combined, even though the trend is reversed when each stone group is considered separately. These reversals will happen when causal effects are small in relation to the effect of confounding variables, as stone size is a much more important factor in determining outcomes here than the actual treatment.

Pearl's Bayesian Networks approach would be able to discover and control for the effects of such confoundedness in special circumstances where the effect of any confounding variable(s) can be measured. However, if the confounding effects are not measured in covariates, or if there are more complicated effects from multiple variables being confounded, then Pearl warns that this approach may not be successful in removing bias in estimates of causal effects.[39] Like Bayesian Networks, RELR's outcome score matching method would invalidate spurious effects due to Simpson's Paradox that are the result of sampling imbalances and confounding effects, as long as the confounding factors are measured as manifest or latent factors which is more likely with RELR's high-dimensional modeling. However, Explicit RELR's data mining may miss small causal effects due to Simpson's Paradox, as it may not see an effect that is small and canceled by an opposing or larger confounding and spurious effect. Therefore, the net is that RELR's automatic machine learning methodology is much more likely to find larger causal effects that are unlikely to fall victim to Simpson's Paradox, as it will omit spurious effects that do not pass the quasi-experimental testing and small causal effects that are obscured in the Explicit RELR data mining by spurious effects or larger causal effects. On the other hand, if a researcher had a causal hypothesis that a small effect was present, RELR's outcome score matching could be used to test this hypothesis and control for any measured confounding features, but this would no longer be an automatic machine learning method.

Model identification issues related to unmeasured causal variables are another critical problem in Bayesian Networks. In particular, if causal variables are not measured, then the model may be biased to lack important causal effects and overstate or incorrectly specify those effects that are in the model. In some cases, latent variables constructed from measured variables may substitute for actual causal variables, but these latent variables must be constructed from covariates that are not confounded. For this reason, Pearl warns that proper model identification cannot be generally guaranteed. The formalism that Pearl introduces is helpful to understand in a theoretical sense how some causal effects could be properly identified while others cannot be identified.

Another approach to model Bayesian Networks is logistic regression modeling.[40] The logistic regression Bayesian Network does not require the local conditional independence assumptions of the noisy-OR model. In this logistic regression network, all nodes such as age and head injury in the above probable Alzheimer's model are simply modeled as input variables in the logistic regression; interaction effects also can be modeled. The only limitation is that all outcome variables must be discrete in logistic regression, as continuous dependent variables are not allowed. Any continuous variable can be broken into discrete ordered categories though, so this is a general approach to constructing a Bayesian Network. Thus, in cases where these effects give better predictions without forcing parameters consistent with local conditional independence, the logistic regression Bayesian Network would perform better, assuming the same structure. Another advantage proposed for logistic regression over noisy-OR is that regularization is possible to control overfitting.[41] There are a number of studies that compare standard logistic regression models to noisy-OR Bayesian Network models, although the structures are different, so these results are difficult to interpret for this and other reasons.[42] Some of these studies give a slight benefit to noisy-OR Bayesian Networks over standard logistic regression.[43,44] But, the interpretation also remains difficult because regularization was not performed in the logistic regression.

Along with honoring the time-dependent characteristic that causes must precede effects, Pearl suggests that a causal model needs to have two other very important properties. These properties are found in Explicit RELR models, and they are parsimony and stability. Given that any set of input features that converges on a node in Bayesian Networks can be built with standard or regularized logistic regression, it also would be straightforward to build such networks with Explicit RELR. While Explicit RELR requires feedback in its model training, feedback is not required once the model is learned so the directed acyclic feed forward structure that characterizes Bayesian Networks could characterize the Explicit RELR model then. The biggest advantage may be that Explicit RELR is not up against NP-hard limitations in its selection of an optimal solution, as it in fact is not computing the most minimal conditionally independent solution. That is, Explicit RELR does not impose any restrictive conditional independence assumptions, so the NP-hard limitations related to learning network structure do not apply. Because of this, Explicit RELR solutions will be likely to be optimal in the sense of its *RLL* objective function at least in independent representative test samples. Very complex networks can be constructed simply by layering Explicit RELR models upon one another where the output from any given node in a network can join other features and become an input feature for the next node which becomes an input feature for the next node and so on.

The combined action of Explicit and Implicit RELR that produces the matched sample in RELR's conditional logistic regression causal testing process is an automated process that does not require causal graphical networks methods with human judgment at any step in the process. In fact, it is simply not possible for a human to consider a complex causal graphic relationship for every possible causal candidate feature along with all possible interaction and nonlinear effects due to the extremely high dimensions that often exist. In any case, human judgment and creativity certainly will be involved in interpreting causal features that are automatically discovered through RELR's mechanical process, especially with respect to underlying latent mechanisms that such features represent.

There may be simple causal reasoning scenarios, for example with young children, which reduce to a clock-like Bayesian Network with conditionally independent features, but even then the observation learning can be understood as a matching quasi-experiment.[45] In such experiments, children are likely to attribute causality to an object that turns off music when placed on a music box by itself, and not attribute causality to objects that do not turn off the music by themselves. The causal reasoning is essentially one of the comparing outcomes across conditions where an object is varied on the music box to be "by itself" or "with other objects" but all other factors are matched across the situations. Random assignment of these conditions is not what is critical for the brain's causal learning here. Instead, the fact that all other factors are matched across these conditions is what is critical. In that sense, the brain's causal reasoning is exactly what occurs in a matching quasi-experiment and is what RELR's causal reasoning mechanism is designed to model. Yet, RELR is able to consider many more potential confounds than the conscious human brain through its high-dimensional modeling, and thus RELR has the potential to be much more powerful at detecting true causal relationships from observation data.

RELR's automatic causal reasoning mechanism is a merger of a number of basic computation mechanisms including Explicit RELR to form stable causal hypotheses, Implicit RELR to form stable high-dimensional models from which to select matched samples, Topsøe distance to form stable matched samples, and conditional logistic regression to test those matched samples in terms of a putative causal effect. These basic mechanisms are themselves all related through their connection to very simple maximum entropy and likelihood computation mechanisms in logistic regression that are stable because of RELR's error modeling. Taken together with the minimal KL divergence used in RELR's sequential online learning and memory updating, these basic computational mechanisms are all related in a fundamental way to information theory. The next chapter begins to explore how many of these same information theory mechanisms could be embodied in neural computations.

CHAPTER 5

Neural Calculus

*"[Neurons] are small, globular, and irregular… [and] supplied with
numerous protoplasmic prolongations [dendrites]. The special character
of these cells is the striking arrangement of their nerve filament [axon],
which arises from the cell body …"*
Santiago Ramón y Cajal, Estructura de los centros nerviosos de las aves, 1888.[1]

Contents

Just as the existence of atoms had been doubted through much of the nineteenth century, most scientists also did not believe in neurons in those years. Prior to the work of Cajal, the prevailing view promoted by Golgi was that the nervous system was continuously connected without discrete components through a mesh of fibers called the reticulum.[2] While Cajal's discovery of neurons as discrete units that are connected through synapses has had profound effects in medical and pharmaceutical applications, it really has not yet had much effect in everyday predictive analytic applications of machine learning. Most of today's most widely used machine learning methods to model and predict human behavior outcomes, including the popular multilayer back propagation supervised artificial neural networks, do not seem to have a strong connection to real neurons especially given what we now know about how neurons work. This is similar to how alchemy had no connection to atoms.

Leibniz's goal to find a *Calculus Ratiocinator* has turned out to be a very difficult problem due to all the possible arbitrary ways that analytics can be generated to predict and explain complex human behavior-related outcomes. We could rely upon pure statistical theory to avoid arbitrary features. But our solution would be just a theoretical mathematical solution that may not reflect any natural mechanism that actually produces cognition and

Calculus of Thought
http://dx.doi.org/10.1016/B978-0-12-410407-5.00005-2

behavior. Statistical theory also has limitations as underlying assumptions may not be correct. But if the supposed nonarbitrary statistical properties are also seen in neurons, then there will be better chances of being on the right track. RELR does provide a feasible solution grounded in statistical theory that that has important computational features that are also seen in neurons.

Obviously, RELR is not the first attempt to base a predictive analytic method upon neuroscience as the artificial neural networks that were introduced in the 1980s made that attempt. Also, RELR is not the first proposal that the neuron is essentially a probability estimation computer constantly predicting its own maximum probability output.[3] What then is different about RELR's modeling of computations at the neural level? First and most importantly, the error probability modeling that is the basis of RELR is what is most different from standard artificial neural network methods. This ability to deal effectively with error is a very important feature that is expected also to occur within neurons. This is because accurate small sample learning with high-dimensional multicollinear input effects also must happen in neurons.[4,5] Second, much more is now known about neural computations than when artificial neural networks were introduced 30 years ago, so it will be seen that other aspects of RELR related to how it handles the computation of interaction and nonlinear effects also seem closer to how real neurons work. Third, Explicit and Implicit RELR are general enough to model both explicit and implicit learning. Fourth, the RELR sequential online learning process generates prior weights from historical learning that can be interpreted as memory weights in ways that that memory weights in neurons also may contribute to learning. As more is understood about neural computation, it may become obvious that RELR has serious shortcomings. In any event, RELR does seem more similar to real neural computation than the artificial neural network methods from the 1980s.

1. RELR AS A NEURAL COMPUTATIONAL MODEL

Neural processes that drive coordinated movement, attention, perception, reasoning, and intelligent behavior result from learning and memory. Not all these memory processes are necessarily conscious and explicit, as much of what happens in the brain only implicitly supports rather than explicitly reflects the conscious processes of thinking. Yet, it is assumed that all these processes must result from calculations directly performed by neurons, or related synchronized population ensemble effects that are manifested through the field surrounding neurons.[6]

Figure 2.1 provides a sketch of neurons drawn by Santiago Ramón y Cajal in 1899. While there are many different types of neurons with different firing properties, all neurons produce graded signals that either directly or indirectly can cause binary spiking signals.[7] So the computational mechanism that leads from graded signals to binary impulses should be the basis of learning and memory. The most fundamental aspect of a cognitive calculus then really is the explanation of how this fundamental neural computation mechanism works to produce learning and memory (Fig. 5.1).

RELR is a mechanism for the computation of memory weights that ultimately determine binary on–off spiking responses. In real neurons, this needs to be done so that reliable and stable learning can occur quickly across the large number of synaptic inputs and associated interaction and nonlinear

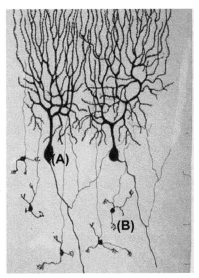

Figure 5.1 Drawing of neurons in pigeon cerebellum by Santiago Ramón y Cajal, 1899; *Instituto Cajal*, Madrid, Spain. Purkinje cells are exemplified by (A) and granule cells by (B). The dendrites of the Purkinje cells are easily seen at the top as abundant tree-like structures. The cell bodies are also obvious as the oval-shaped structures in the middle, and axons are seen exiting these cell bodies.[8] Neural electric information flows from dendrites to cell bodies to axons. Gaps where information jumps to neighboring neurons are called *synapses* which are usually bridged at axonal terminals through chemical messengers called *neurotransmitters*, although electrical tight junction synapses also exist. (For color version of this figure, the reader is referred to the online version of this book.)

effects as would be measured at the point where the soma meets the axon, which is the initial zone within the neuron where axonal spiking occurs. So, as a neural computational model, RELR is a mechanism for how binary on–off spiking responses would be learned as a function of these many synaptic input signals and their corresponding interaction and nonlinear effects. Figure 2.2 describes all steps in the basic RELR computational process that are seen in both Implicit and Explicit RELR learning.

The weighted feature effects in the RELR model that are summed in step 2 in Fig. 2.2 are akin to graded potential effects at the soma in a real neuron where the weightings of step 1 are prior weights. Each weighted feature effect arises directly either through synaptic inputs or through their associated interaction and nonlinear effects as determined by neural dynamics. In real neurons, an axon may have terminal branches to thousands of synapses,[9] and many of these could terminate on the same target neuron. So it would be reasonable to believe that each neuron receives many synaptic inputs from the same presynaptic neuron. If an input feature is defined as the sum of graded potentials that arise at the soma due to a specific presynaptic neuron but with some possible time variation across the many input synapses from this specific neuron, then it is clear that an input feature could arise due to either a binary or continuous variable. Not all input connections between neurons are across chemical synapses as electric synapses exists across tight gap junctions. Such signals may not be expected to be binary, and instead could be continuously valued. Thus, each input feature may be regarded as a main effect independent variable that can be either binary or numeric in form in the sense of classic regression modeling.[10] These input features also produce interaction and nonlinear effects due to neural dynamic effects within dendrites, and the interaction and nonlinear effects also may be regarded as independent variables in a regression model. The exact form of the weighting function in step 3 is determined by the RELR computations which have learned to maximize entropy historically given inputs from presynaptic neurons. Step 4 is the binary spiking that initially occurs at the soma and is transported down the axon and communicates with target neurons (Fig. 5.2).

The learning weights computed through the RELR method are automatically adjusted to account for the probability of error in the input features. This error modeling is based in part upon Hebbian spike-timing plasticity mechanisms where it is assumed that connected groups of neurons with correlated activity should wire together.[12] Correlation here is meant in a general sense to imply both positive and negative correlation. This

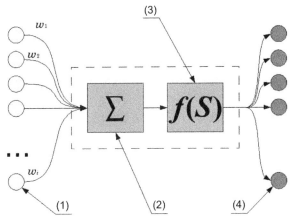

Figure 5.2 The RELR learning method is a maximum entropy/maximum likelihood process that is a form of logistic regression (Appendix A1). This is a model for neural computations as follows: (1) a set of prior memory weights *w* are learned to determine the binary spiking probability of a neuron at the soma through the reduced error logistic regression method that includes error modeling; (2) the independent variables, including presynaptic input main effects and their interaction and nonlinear effects, are multiplied by each respective prior memory weight and pertinent update weights and summed (Appendix A4) the exact form of the function *f*(*S*) that is summed at the soma is determined by independent variables and their learning history; and (4) spiking signals at the soma may be produced according to the dynamics of the simple model (Section 3) or a more complex dynamic model. Spiking signals in axons represent the final binary output and the prior memory weights in step 1 are then updated based upon the posterior weights which are the sum of the old prior and update weights for each feature (Appendix A4). The degree to which the update weights are nonzero is a function of whether or not binary spiking occurs over a period of time, and whether that spiking deviates from historical spiking patterns. Permission granted for use.[11] (For color version of this figure, the reader is referred to the online version of this book.)

assumption is implemented in RELR so that main input features and/or their interaction or nonlinear features that are relatively uncorrelated with binary outcomes are prevented from influencing binary outcome learning. This is accomplished through the RELR feature reduction mechanism that is based upon the magnitude of *t* values that reflect correlations between independent variable features and outcomes as reviewed in previous chapters. Thus, the noisiest features are dropped so that they do not have any influence on binary outcome learning.

The RELR error modeling has also been detailed in previous chapters. This RELR error modeling process is expected to be similar to the actual physical process that occurs at the soma. In the summation process at the soma, probabilities of positive and negative error in graded potentials that determine axonal spiking should be equal without any inherent bias just as is assumed to occur in RELR.[13] That is, the probabilities of positive and negative error are assumed to be equal across all constraints on features in the RELR maximum entropy/maximum likelihood optimization. A further symmetrical assumption is that there is no inherent bias that favors the probability of greater positive or negative error in linear/cubic effects versus quadratic/quartic effects in these feature constraints (Appendix A1). These symmetrical error probability constraints effectively assume that all error is likely to cancel out when the many weighted input effects combine together as happens in the summation step 2 in Fig. 2.2 in the RELR model. Again, in real neurons, it is expected that positive and negative errors also should cancel on average without bias as weighted graded potentials are summed together in a final binary decision step at the soma that determines whether or not the axon sends a spiking signal.

2. RELR AND NEURAL DYNAMICS

Neural dynamics allow for the possibility of nonlinear and interaction effects in neural computation and determine how neurons make firing decisions. Neural dynamics characterize the electric and chemical gradients and flows of ions that cause individual neural components to produce passive and/or spike signals. Step 4 in Fig. 2.2 is the binary spiking rule that determines how binary outputs are ultimately produced based upon the weighted sum of effects. A simple static binary decision rule based upon either the KS-statistic or the predefined probability threshold is usually used in real-world logistic regression models including RELR. Such a static binary decision rule is akin to an "integrate and fire" neural decision where all weighted effects sum together and cause a binary "yes" decision when that sum is greater than a threshold. However, many neurons may use much more sophisticated spiking rules than the simple integrate and fire rule. These more sophisticated rules may be dynamically learned and may change during important stages of neurodevelopment.

Hodgkin and Huxley first produced a reasonable quantitative model for spike generation in the squid axon in the 1950s, but the field of neural dynamics has evolved considerably since then and especially in the past 20 years. When I was a graduate student studying cognitive neuroscience in the 1980s, I was taught the prevailing view based upon the Hodgkin–Huxley model that all neurons used an integrate and fire mechanism at the soma that caused axons alone to transmit all-or-none spike signals based upon a defined threshold. This Hodgkin–Huxley model is now taught to be just a special case of how neurons actually work.

A much more general quantitative theory for neural dynamics has been succinctly summarized in a model proposed by Eugene Izhikevich that he calls the simple model (Fig. 5.3).[15] This Izhikevich simple model captures known facts that were missed in the Hodgkin–Huxley model including that many dendrites can also produce spikes, many neurons do not have clearly defined firing thresholds, many neurons do not fire just one spike but instead fire bursts of spikes, and many neurons do not simply integrate and fire but instead may resonate and fire. The beauty of this simple model is that it can reproduce realistic neural dynamics including bursting oscillations in a wide variety of neurons with few parameters in a system involving just two linear differential equations.

The v variable in Fig. 5.3 reflects the fast postsynaptic graded potentials that are evoked by the synaptic input current denoted by I; the u variable

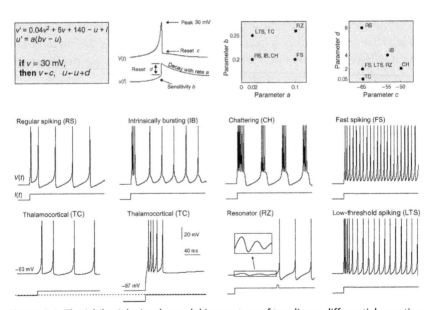

Figure 5.3 The Izhikevich simple model is a system of two linear differential equations that reflect the time derivatives (v' and u') of a fast membrane potential variable v and a slow ionic current variable u. The parameters a, b, c, and d can be measured in real neurons. The parameter I is the injected sum of synaptic input current amplitude. By varying the parameters in the model, the simple model can accurately describe the dynamic behavior of a very large number of neurons, some of which are shown here.[14] (For color version of this figure, the reader is referred to the online version of this book.)

reflects a slower membrane current recovery variable that likely would be intrinsic to specific types of neurons and synapses and describes the recovery of membrane currents back to equilibrium. The v variable reflects inputs in the form of graded potentials that originally arise at synapses, whereas the u variable(s) is an intrinsic variable that is similar to what statistical modelers would call an intercept in that it controls the prior probability of a response although it exhibits slow time varying dynamics. The parameters, a, b, c, and d are the properties that can be measured in real neurons. Given these measured parameters, the dynamic firing properties are then entirely determined by the coefficient weights of the u and v variables in these equations.

The simple model in Fig. 5.3 is just an estimate of the simplest model needed to reproduce neural dynamics in a wide variety of neurons, but some real neurons could be adequately modeled with a zero weighting of the quadratic v^2 term or linear v term. In fact, the simple model allows for a more general model that would allow for variation in the weighting of the v or v^2 term, or for terms to be dropped.[16] This simple model is apparently very sensitive to the choice of the after spike resetting cutoff parameter and can be unwieldy in hardware implementations for that reason. Alternative models that employ either a quartic v^4 or an exponential term instead of v^2 may not have this problem,[17] although adjustments can be made to the parameters of the simple model to avoid such instability.[18] In any case, models that have a quartic term in addition to a linear term rather than an exponential term exhibit sustained subthreshold oscillation, which is believed to a better model for the dynamics of some types of neurons.[19] Thus, this characterization of neural dynamics continues to be an evolving field, but it does seem to be the case that one or more nonlinear terms, such as v^2 or v^4 may need to be used as potential basis features in these equations. It is also likely that real neurons are more complex than these simple possibilities, as a predominant v^3 term also could be a more realistic possibility in at least some neurons in what is known as the Fitzhugh–Nagumo model.[20] The reality is that all neurons are different,[21] so such diversity would be very much expected across real neurons.

These models like the Izhikevich simple model are quite general and could reflect dynamics within dendrites, at the soma, and along the axon. The dendritic tree of a neuron often has extensive branching, so this allows for signals originating from diverse presynaptic inputs to interact at junctions in this branching. To simplify simulations, dendritic computations can be modeled as separate computations that employ this same simple model within each small dendritic compartment.[22] These multicompartment models typically treat the change in graded potentials across time or v' in this simple model as a sum of all graded potentials v in neighboring compartments.[23] In this case, the equation that describes v' in a given compartment then becomes a complex sum of similar expressions across all preceding compartments[24] but with many different graded potential variables and can be represented by a multiple regression equation.[25] So the v variable in this equation would be more realistically depicted as a vector with each element in the vector reflecting a feature given by a different presynaptic neuron, and with the coefficient weight also allowed to vary with each element and with associated nonlinear effects. Thus, this equation describing the change in soma graded potentials per unit change in time or v' becomes a complex regression equation with many graded potential effects determining the probability of axonal spiking and this could be represented in a multiple logistic regression model.

An even more realistic model might allow for many independent interaction effects between separate presynaptic inputs, as this would be consistent with discussions about the nonlinear and coincident interaction computational properties of dendrites.[26] A large number of main effect variables in the vector v and associated nonlinear effects such as v^2, v^3, or v^4 along with a much larger number of associated interaction effects at junctions between dendritic branches are possible with many spatially diverse dendritic compartments. These many inputs and associated interaction and nonlinear effects that would arise are interpreted to be independent variables in the RELR statistical model. Thus, unlike classic neural network models which learn synaptic weights across synaptic inputs, RELR as a model of individual neural computation might be interpreted to be a model of the learning of the weights that arise in a complex dynamic model at the region of the soma that ultimately determines axonal spiking.

Technically, graded potential effects at the soma are not directly related to synaptic weights because they are determined by the passive cable and active dendritic spiking properties of the neuron that reflect interaction and nonlinear dynamic effects within neurons. However, in a neural computation model, RELR regression coefficient weights could still be viewed as memory weights that are indirectly affected by synaptic learning. Also, the RELR memory weights would directly impact the probability of neural spiking in such a model, but they just would not be synaptic weights. Understandably, multicompartment neural dynamic models wish to separate synaptic weights from neural dynamic effects.[27] But this combination of all such weight effects into a potentially high-dimensional multiple logistic regression model that determines spiking probability at the soma has its advantages in that it becomes an aggregate representation that is tractable analytically as modeled in the RELR method, whereas more complex multicompartment models only can be understood through simulations. This also allows one to interpret the RELR computations in terms of putative learned computations performed within neurons that determine axonal binary spiking at the soma. Thus, RELR is clearly different from traditional neural network models that attempt to model synaptic weight learning directly,[28,29] as RELR models the learning of weights that arise at the soma through a combination of neural dynamic and synaptic learning effects.

So, whereas the Izhikevich simple model and similar models are describing how neural dynamics are determined through realistic weight parameters, RELR as a neural model simply characterizes how the memory weights in these models would arise through learning, where learning is defined very broadly to mean genetic and environmental historical learning. Thus, the coefficients of the fast membrane potentials vector v and all associated nonlinear and interaction effects in neural dynamics models are the very same posterior weight parameters that the RELR method learns automatically through the maximum entropy subject to linear constraints method. Prior genetic or environmental weighting can be interpreted to be prior weights that can be unique to each feature (Appendix A4), which functions in a similar way to the u slow ionic current variable in the Izhikevich simple model in that it can be slowly varying in time until it abruptly is updated with binary spiking. RELR's learning results in optimal posterior coefficient memory weights that maximize entropy subject to all independent variable constraints. The independent variable features may be highly multicollinear features that include all main input effects and resultant interaction and nonlinear effects in what could be a massively highdimensional statistical model as in Implicit RELR, or a parsimonious feature selection model that selects very specific main or interaction or nonlinear effects through Explicit RELR.[30]

3. SMALL SAMPLES IN NEURAL LEARNING

Through cached, online small sample update learning in working memory, the brain seems to be able to learn novel complex input patterns very quickly. Some medial temporal lobe neurons actually show single-trial learning in recognition memory tasks,[31] although a single learning trial may be associated with numerous bursts of many binary responses in a single neuron, and rehearsal of the very same binary spiking input pattern presumably also may take place in working memory. Yet, the total number of binary responses still must be relatively very small for stable learning in a single neuron with a complex input pattern such as in recognition memory discrimination. In fact, humans can learn the meaning of a new word in the course of seconds.[32,33] The maximal firing rate of neurons is expected to be roughly 1000 spikes/s, but the average firing is expected to be much lower. So, within hundreds of binary responses, typical neurons involved in higher level cognitive functions such as recognition memory or semantic memory may begin to show learning.

Just as is supposed to be happening in neural learning, RELR is a computational engine designed to learn the probability of binary outcomes given a small sample of training observations which also potentially has a very large number of independent features. Unlike standard predictive analytic methods, RELR estimates the probability of error and the probability of these binary outcomes. For this reason, RELR is able to remove error and produce solutions that are much more stable and much less prone to error with small observation samples and/or large numbers of multicollinear independent features.

Some process must be going on in neural learning that is quite different from standard regression methods because neurons must have an ability to learn with relatively small numbers of spiking responses compared to a huge number of inputs and associated nonlinear and interaction effects. For example, a typical neuron receives input from 10,000 neurons.[34] Even if there is much overlap in these binary inputs such that they essentially code a smaller number of multicollinear continuously-valued features, there still would be a very large number of candidate features. With only 1000 independent input features from separate neurons, there would be the same number of 1000 independent linear input variables in a regression model that are likely to be highly multicollinear. With two-way interactions between linear variables, this would yield 501,000 total effects to consider including the interaction effects.[35] If nonlinear effects up to a fourth-order polynomial are allowed, this would give over 2,000,000 largely multicollinear effects in a regression model.[36] The traditional 10:1 rule of thumb that stated that one needs 10 response observations per linear variable does not apply here due to the nonlinear and interaction effects. So, a more general Dahlquist and Björk rule[37] which requires a vastly greater number of observations is needed. This more general rule of thumb would require a million binary outcome observations with only 2000 independent variable effects and a thousand million such observations with only 60,000 variable effects that include nonlinear terms such as interaction effects. Even with only 501,000 independent variable effect terms as in this original example which only includes two way interaction effects, such processing would require >348,600 hours,[38] which would be close to 40 years for a single neuron to learn the input response pattern.

So obviously neurons cannot learn in accordance with standard regression methods because those methods require such huge numbers of observations to avoid multicollinearity and overfitting error. There are some newer methods such as Random Forest, L2 regularization or Ridge Regression, L1 regularization or LASSO, and Support Vector Machines that deal with this error problem through averaging of different solutions or through regularization, which can work to smooth away at least some error. But these methods still have fairly unstable parameters and overfitting error with small samples, along with arbitrary parameters to control the degree of smoothing or averaging.[39,40] There are also multilayer supervised neural network methods that incorporate similar regularization as Ridge Regression[41] and also can show better prediction in small samples. But they also have similar problems as Ridge Regression with unstable parameters and overfitting in small samples, along with an arbitrary smoothing parameter that will change when a different arbitrary cross–validation procedure is used to tune this regularization parameter. Naïve Bayes is known to give relatively accurate prediction and stable parameters with very small training samples.[42] But Naïve Bayes cannot see interactions and is susceptible to overfitting. In general, all the traditional methods seem to require too much training data to reflect the neural learning method or have other problems. Another possibility is that massively parallel groups of neurons could use many independent methods or modeling parameters and then average their results so that error in small sample learning is averaged out. This would work if the massively parallel learning used a significant number of independent methods that give reasonable importance sampling as can be done in effective stacked ensemble model learning. However, it is hard to understand how the brain evolved or learned such a complex process as stacked ensemble learning. As reviewed in the initial chapter, even though implicit memory models do seem similar to ensemble models once they are learned, these stacked ensemble models do not allow the automatic learning that would be expected in the brain's implicit learning.

A much simpler explanation would be that the basic neural calculus has evolved to learn most probable responses that are very effective at removing error probabilities. A common theme in all previous attempts to handle error in predictive analytic methods is that error is not viewed from the perspective of a probability model. As reviewed in the initial chapter, various regularization methods like LASSO and Ridge do not estimate the error in terms of probability of error that sum to one for each independent variable, as the L1 and L2 penalty terms are not probability measures but are instead arbitrarily weighted nonprobability functions. By contrast, RELR estimates the probability of all outcome and error events given the input features. Because of this accurate error probability modeling, RELR automatically handles error and produces small sample learning with high-dimensional multicollinear inputs.

As a neural model, the RELR error model is assumed to be the natural outcome of a maximum entropy process where graded potential effects sum at the soma so that the probability of positive and negative errors cancel. The Hebbian learning principle is one of the most widely applied ideas in neural learning; it is that "neurons that fire together wire together".[43] In a broader sense where both positive and negative correlation drives learning, spike-timing effects in this RELR process could determine that input effects that are poorly correlated with the spiking output of a neuron would not influence the learning mechanism just as RELR's feature reduction removes small magnitude effects. Moreover, RELR's implicit and explicit selection methods have a strong preference for feature effects that are more positively or negatively correlated with binary output spiking patterns just as would occur with spike-timing dependent learning. Unlike stacked ensemble learning, the Explicit and Implicit RELR maximum probability methods could be embodied in individual neurons. Unlike all these other methods that require extra model tuning based upon arbitrary cross-validation samples to avoid overfitting, RELR's learning mechanisms are completely automated and entirely based upon the training sample. In addition, RELR's sequential Bayesian online learning method could directly occur in sequential phases of cached, small sample "working memory" update episodes. Taken as a whole, these computational properties of RELR are the characteristics that might be expected in real neurons in the brain that are the basis of learning and memory.

4. WHAT ABOUT ARTIFICIAL NEURAL NETWORKS?

These suggestions that RELR could be a reasonable model for the calculus that determines learning and memory in real neurons may cause some to remember the claims made by artificial neural network modelers in the early 1980s. I recall being at a packed conference at UC-Irvine in the mid-1980s when there was great excitement about the newer artificial neural networks. The single-layer perceptron proposed in the 1950s had been the original neural network method.[44] As Marvin Minsky famously showed, this single-layer perceptron could not learn Exclusive OR and Exclusive XOR predictive relationships, which are what statisticians would call interactions.[45] So the single-layer perceptron had classifier properties that are about the same as would be expected from a standard logistic regression model without interactions. Much of the excitement in the 1980s concerned the fact that new multinode and multilayer artificial neural networks finally could model interactions effects.[46] In fact, there was unbelievable optimism back then that these new artificial neural networks were reflections of the brain's cognitive computation mechanisms. Yet, very few of the promises from the great hype of the 1980s have been realized.

The neural networks that grew out of the 1980s were actually a large number of different methods but one of the most popular methods was a supervised neural network that most often employs the back-propagation method across a multilayer network of artificial neurons that includes a hidden layer. These multilayer perceptrons are designed to minimize the difference between the target or supervising signal and the output of the system. Because of the hidden layer, these multilayer perceptrons suffer because they are black boxes which are not transparent in terms of understanding how they work. Another problem is that there has never been general empirical support for supervised back-propagation learning across a network of neurons in the brain. However, there is now some evidence that a back-propagation learning mechanism could exist within individual neurons. This within-neuron back-propagation mechanism is thought to be involved in spike-timing-dependent learning processes.[47] There has been a fairly recent proposal that a multilayer perceptron using the 1980s back-propagation idea could involve a dendritic hidden layer within neurons and explain neural learning computations.[48] Yet, a number of questions have been raised as to whether this model's dendritic hidden layer computations could actually happen in real neurons.[49] Even if the new knowledge about within-neuron learning mechanisms could be aligned with these classic supervised artificial neural network methods, these multilayer back-propagation methods from the 1980s have a more fundamental flaw that was reviewed in the last section. Even when they employ regularization to attempt to smooth away error, parameters are still too

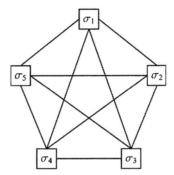

Figure 5.4 A schematic of a fully connected Hopfield neural network.[50]

unstable and tend to overfit too much in small sample sizes and with large numbers of multicollinear features to be a reasonable model for neural learning.

Another well-known method from the 1980s was the Hopfield neural network depicted in Figure 5.4.[51] In a fully connected Hopfield neural network, each node is considered to be a neuron which is connected as an input to all other neurons and which receives all of its inputs from all other neurons. The Hopfield neural network is a recursive network that requires feedback in order to compute its weights in a model of associative memory in that each input learns to be associated with all others in network. Hopfield networks suffer from the problem that its solutions are not ensured to be globally optimal, and instead are often local minima. Hopfield networks are closely related to Boltzmann machines which are another type neural network method with roots in the 1980s. The major difference is that the Boltzmann machine's optimization is stochastic and can use simulated annealing which might help it escape local minima,[52] whereas the Hopfield network employs a deterministic method.[53]

The original Boltzmann machine suffered from difficulties in converging in reasonable periods of time with more complex problems, but Restricted Boltzmann machines are now used which simplify the connection architecture in the original Boltzmann machine to produce much faster computation although hidden units are still usually employed, and back propagation may be used in deep learning tasks.[54] New formulations also have been proposed to help with some of the more severe problems in Hopfield networks.[55] Yet, this can change the basic simple functionality and interpretation by imposing learning mechanisms that are no longer based upon simple cross-products to increase network capacity.[56]

RELR is similar to Restricted Boltzmann machines, Hopfield networks, and even the multilayer back-propagation perceptron in the superficial sense that it is also an optimization method. But RELR also takes into account errors-in-variables. For this reason, RELR may be more likely to find estimates of global maxima that generalize well to new observation samples. In this sense, RELR's error modeling also may be similar to what occurs in real neurons in that it allows very high-dimensional modeling in small samples that avoid unstable solutions and overfitting. Another advantage is that RELR allows online sequential learning, along with causal learning, in completely automated machine learning computations that are at least putatively similar to neural computations in ways that ultimately could be tested empirically.

Other artificial neural network methods developed in the 1980s were unsupervised neural networks, which can be used in dimension reduction tasks. Neural networks designed to categorize visual images like dogs and cats into groups of objects often have an early unsupervised processing stage that may be almost equivalent to K-mean clustering, as exemplified in the work of Riesenhuber and Poggio.[57] This dimension reduction stage reduces extremely high-dimensional input data into a manageable smaller set of input features for later supervised neural network processing.[58] Another well-known unsupervised neural network approach from the 1980s is the Self-Organizing Map method which is essentially a form of nonlinear Principal Component Analysis.[59]

A problem with the supervised and unsupervised neural network dichotomy from the 1980s is the nagging question about whether the brain actually performs supervised versus unsupervised learning. In fact, one very influential view in today's neuroscience is that the classic supervised versus unsupervised neural learning dichotomy from the 1980s is very artificial and is not the way that the brain actually works.[60] This argument is that the relationship between neurons and their input effects should be interactive rather than supervised or unsupervised and thus may depend upon resonance between binary neural spiking and local field potentials. In accord with this alternative view, one artificial neural learning method that was first introduced in the 1980s is fundamentally different from methods that assume that the brain must have separate supervised and unsupervised learning mechanisms. This is the neural Darwinism approach popularized by Nobel Laureate Gerald Edelman and other prominent scientists.[61–63]

Neural Darwinism assumes that neurons learn by selecting input patterns of activity, or neuronal groups. This is assumed to occur as neural spiking output patterns resonate with input effects in a natural selection mechanism that is akin to the mechanism at work in the evolution of species. With this emphasis on spike-timing-dependent plasticity as a fundamental neural learning mechanism and given the evidence for spike-timing plasticity mechanisms reviewed in the next chapter, neural Darwinism and this theory of neuronal group selection may be closer to how neurons actually work than all the other neural network methods from the 1980s.[64] Neural Darwinism is similar to genetic programming in that it is inspired by genetic principles of natural selection, but it also differs from genetic programming in respect to replication in that it is much more of a random search. For reasons connected to this random search, neural Darwinism is not without critics,[65] especially in reference to a poor ability to produce neural learning that is capable of self-replication. Replication is one of the hallmark genetic actions as observed in DNA mechanisms, but it is argued that neural Darwin mechanisms would produce too many false positive and negative errors to allow neural learning patterns to self-reproduce.[66] While other evolution inspired methods like genetic programming can lead to replication, they also require arbitrary user-dependent parameters that determine mutation and cross-over. For this reason, genetic programming models can have the same flaws related to arbitrary user parameters as the other standard predictive analytic methods.

Neural Darwinism has had success in the neuroscience community as evidenced by its influence in ongoing projects to simulate learning in large networks of artificial neurons.[67,68] Therefore any improved method designed to model neural computation likely will need to consider those aspects of neural Darwinism that have been most effective. Like neural Darwinism, RELR's implicit and explicit learning methods are also each not just passive but rather interactive as the ultimate mechanism of these RELR methods is to select optimal and most probable weights based upon how input effects resonate with spiking outputs. Like neural Darwinism, RELR avoids the artificial dichotomy between supervised and unsupervised neural learning. Therefore, unlike supervised neural networks that require separate unsupervised neural networks to perform dimension reduction preprocessing, RELR can model dimension reduction built into individual neurons based upon magnitude of correlations between input effects and spiking output. In this way, RELR could be a model for the brain's neural computation through resonance between local field potential input effects and binary spiking without a need for separate unsupervised networks to perform dimension reduction preprocessing.

Unlike neural Darwinism, RELR allows interpretable and understandable solutions in terms of parsimonious feature selections that are stable and do not depend upon a random search that causes variability in selected features across independent representative training samples. This stability in RELR's learning is due to accurate error modeling which allows most probable features easily to be discovered. RELR is also able to produce output patterns that perfectly replicate input patterns, so under balanced sample zero-intercept conditions it will avoid the quasi-complete separation convergence problems that plague standard logistic regression, and RELR never will show complete separation problems due to its error modeling. However, RELR is not designed to model genetic mechanisms but is instead designed to model neural computations so any analogy to evolutionary and genetic mechanisms might ultimately break down.

RELR's advantage over a random search is not just in stability and error avoidance. This is because RELR's gradient ascent directly optimizes and is thus substantially faster than the slow, random neural Darwin search. In fact, RELR yields identical solutions as the maximum entropy subject to linear constraints method and has a physical and neurophysiological interpretation. That is, RELR's fundamental maximum entropy-based learning mechanism can be considered to reflect a basic physical mechanism that necessarily must be embodied in neurons. The tendency to find physical solutions that maximize entropy subject to all other available constraints is assumed to be a basic physical law, the Second Law of Thermodynamics. So the RELR learning mechanisms in real neurons would be expected to obey the same basic physical maximum entropy mechanism that has constrained neural dynamics from Hodgkin and Huxley onward.[69]

To the extent that it might have any similarity to neural computation, RELR should not be thought of as a neural network computation model reflecting specific connections between individual neurons. Instead, any modeled neural computation might be thought to represent the modal activity of an ensemble of similar neurons where the variability in posterior memory weights is given by the standard error of RELR's posterior regression parameters. In fact, a RELR neural ensemble may be itself built from inputs that are also ensembles of neurons. In this neural ensemble, each input and output should be viewed as a collection of neurons which can take on a more continuous range of outputs than binary signals, so the probability of a RELR binary output response given the input features would be a better description of the ensemble output than the actual predicted binary output response. This probabilistic description accords with the idea that RELR as a neural computation model is consistent with local field properties of a neural ensemble, as local field potentials are closely correlated with the time histogram that describes the probability of spiking responses in local neural populations.[70] Although RELR does incorporate typical network concepts of feed forward and feedback signals in its implicit and explicit processes, these concepts also could be associated with field properties of aggregate population signals in the brain which as Nunez and collaborators show are believed to have both traveling and standing wave-like properties.[71] Thus, traveling waves may move only in one direction and be feed forward signals, whereas standing waves arise from traveling waves that feedback to an original source and continually resonate. This resonant feedback results in highly structured periodic disturbances like vibrations in a violin string, and Nunez argues that similar standing waves could be the basis of highly structured momentary oscillating neural ensemble synchrony.

As a model of neural computation, probably the most serious problem with supervised neural networks is that they are supervised. This creates a need for a ghost in the brain to supervise these networks with training labels. RELR can be viewed as a supervised method because it does require training labels which are the spiking binary responses in postsynaptic neurons. Yet, as with neural Darwinism methods, these binary spiking events also could be viewed as arising from an interaction between input feature signals and local field potentials that is guided by natural neural selection mechanisms at work in both implicit and explicit processing. Through this selection, explicit or implicit neural computation could result from the interaction between binary spiking events and the brain's local field potentials with the direction of causality being in either direction, so spikes that are driven by input features may lead to local field reflections of postsynaptic potentials in adjacent neurons or local field potentials may influence binary spiking. In fact, there is new evidence that binary spiking responses in a local neural population can be caused by spatiotemporal patterns in local field potentials.[72] Thus, in RELR's information theory of neural computation, there is no need for a supervising ghost in the brain as both unconscious and conscious cognitive functions may arise through stable interactions between local field potentials and spiking neurons.

Back in the 1980s, nobody had considered that standard supervised multilayer neural networks would have significant overfitting and multicollinearity error problems with high-dimensional data. Back then, there was less concern about the general lack of insight into how these artificial neural network learning mechanisms could be embodied in real neurons. Back in the 1980s, nobody yet thought that extensive dendritic arborization could produce interaction effects within neurons without recourse to multiple layers of processing across neurons. Back then, there was very primitive understanding of spike-timing-dependent plasticity processes involving long-term potentiation and depression, so the idea that resonance is basic to neural learning was not that popular. Back then, all neural-inspired models were implicit learning models that only produced hidden units and black box solutions, as there was not yet a full realization that explicit learning was a fundamentally different process.

The past 30 years have produced enormous advances in neuroscience and the pace of advance seems to be quickening. One area that has produced especially important progress concerns how mechanisms of cognition, learning, and memory may be related to synchronous oscillations in reso-nating groups of neurons in accord with spike-timing plasticity mechanisms. Yet, these principles really have not yet made their way in any impactful sense into machine learning theory and application. The next chapter gives an overview of oscillating synchrony as a fundamental organizing principle of cognitive neural computation.

CHAPTER 6

Oscillating Neural Synchrony

"In the melody, in the high singing principal voice, leading the whole and progressing with unrestrained freedom, in the uninterrupted significant connection of one thought from beginning to end, and expressing a whole, I recognize the highest grade of the Will's objectification, the intellectual life and endeavor of man."

Arthur Schopenhauer, The World as Will and Representation, 1818.[1]

Contents

There is an enormous gap today between knowledge at the individual neural level and ensemble level. The technology simply has not existed to fill this gap with good empirical measurement. Instead, computational modeling methods have attempted to fill this divide, but these insights are quite fallible including those of Reduced Error Logistic Regression (RELR) and its associated information theory unless they are supported by empirical data. Promising new imaging methods are now available that may finally bridge this chasm, and this is a major focus of a new initiative led by the National Institutes of Health (NIH) in the United States called the Brain Activity Map (BAM) initiative. This BAM initiative eventually could be on the same scale and have the same impact as the NIH Human Genome Project. These new imaging methods are based upon nanotechnology and will allow activity from large collections of neurons to be simultaneously recorded.[2]

While the BAM proposal acknowledges that computational modeling approaches will still have a place, all computational models of neural function will be greatly enhanced by these data. In many cases, these data may falsify current proposals that attempt to bridge the enormous gap between individual neural function and populations of neurons including the RELR models reviewed in this book. One may wonder how helpful better data on neural computation and consciousness would be for machine learning applications. Much smarter machines should be possible with better observations of how the massively parallel brain may achieve the intricate timing necessary for accurate and rapid cognitive prediction and explanation. Yet, this knowledge also could be humbling and suggest limitations to cognitive machines designed to simulate these neural ensemble processes.

This chapter introduces the idea of oscillating neural synchrony as a basic organizing principle in ensembles of neurons, as it focuses on abundant research both on an individual neural level and on a population neural level that has concerned how information is rapidly coded and computed in the brain's massive parallel processing. The importance of oscillating neural synchrony in neural computation is unlikely to change as new measurements become available with the BAM initiative. But a much better understanding of how oscillating neural synchrony arises in neural ensembles and how it may reflect cognition and consciousness is a likely possible. Oscillating neural synchrony is the most obvious observation when measurements are taken from the brain's electromagnetic fields through devices known as the electroencephalography (EEG) and the magnetoencephalography (MEG). These measurements are quite noisy and thus require significant averaging or filtering to remove noise. Yet, the resulting signals often have substantial rhythmic and periodic structure that begs comparison to another phenomenon that seems to reflect cognition and consciousness. This is music.

The nineteenth century German philosopher Arthur Schopenhauer saw evidence for intelligence in all forms of matter. To Schopenhauer, there is a natural hierarchical order to intelligence; human thought is simply at the highest level. Schopenhauer suggested that the art form of music is our means to represent and understand this fundamental intelligence process that he called *the Will*. Whereas all the other art forms represent the objective expressions of the Will, music he argued represents the inner or subjective dynamics of the Will. Low-bass tones represent the Will in inorganic forms and high tones represent the Will in organic forms, but Schopenhauer speculated that melody is that aspect of music that represents human thought because it leads with the most informative highest tones and thus connects underlying patterns into a meaningful harmonic whole. Whether or not one accepts the deep mysticism that Schopenhauer borrowed from eastern philosophical traditions, it is hard not to agree that music can represent and move something that is very deep within us. Later sections of this chapter will explore striking similarities between structures that exist in musical forms and the brain's neural signaling at an ensemble level. But first let us understand the concept of neural synchrony and how it is measured through the EEG.

1. THE EEG AND NEURAL SYNCHRONY

The scalp recorded EEG, along with local brain field potentials, are generated by postsynaptic excitatory and inhibitory graded potentials that are synchronized across a large number of neurons in close proximity.[3] While the scalp EEG and local field potentials do not directly reflect the faster spiking responses that represent neural binary signals, the time histogram of spiking responses in a local brain region closely matches the temporal course of local field potentials.[4] Thus, the local field potentials can be thought to reflect nonrandom summation that is roughly the probability of a binary spiking response in a local neural ensemble, whereas the EEG is a cruder and noisier far field reflection of these local field potentials.[5,6]

Neural synchrony as reflected by the EEG often oscillates in very regular rhythms. These EEG patterns are strikingly different across different conscious states. The predominant frequency of the EEG is a rough indicator of the level of consciousness as shown in Fig. 6.1, as faster rhythms imply a more awake brain. However, slower rhythms are often mixed with faster rhythms during intense periods of awake cognitive processing and faster frequencies can be superimposed upon the slowest delta waves during deep sleep.

All important ideas about the neural basis of cognition and consciousness today suggest that it correlates to oscillating neural synchrony. Yet, the greatest degree of neural synchrony is during deep sleep. In fact, the higher amplitude deep sleep delta waves may be described as reflecting a cerebral cortex that is exhibiting the greatest amount of synchrony as reflected by the largest amplitude scalp EEG recordings. How then does one reconcile neural synchrony theories of cognition with the fact that the greatest neural synchrony actually occurs during deep sleep when there seems to be the least cognition and consciousness? While large in amplitude, the slower delta waves that occur during deep sleep are not obviously as structured and periodic as other more rhythmic waves like the alpha rhythm that is shown during the relaxed and wakeful period in Fig. 6.1.

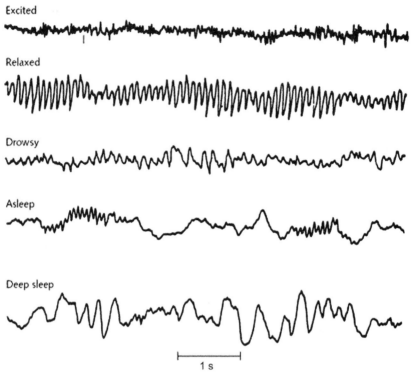

Excited

Relaxed

Drowsy

Asleep

Deep sleep

1 s

Figure 6.1 This early EEG recording depicts the basic levels of consciousness and human cognitive states that are associated with the EEG. Excited conscious states happen during aroused wakefulness and are associated with faster beta- (15–30 Hz) and gamma-EEG frequencies (>30 Hz). Relaxed conscious states occur during reflective wakefulness and are associated with alpha rhythms (8–15 Hz). More drowsy conscious states are associated with theta-range frequencies (4–8 Hz). Sleep and especially deep sleep are associated with slower delta-range frequencies (0–4 Hz). *Source: Penfield and Erickson (1941).*[7]

Thus, the predominant periodicity and rhythmic structure of these oscillating brain fields may be an important indicator of the degree to which brain activity is organized into meaningful cognition. For example, sleep studies show that when theta waves without any obvious rhythmic structure are prominent as occurs in drowsiness and very light sleep, people might report experiencing isolated mental images that are not connected into a coherent story such as seeing an isolated image of a face or hearing the sound of a certain word,[8,9] and subjects might report no memory for any mental activity during deep sleep when delta waves predominate.[10] In contrast, when a sleep research subject reports the experience of a connected train of thought as in dreams or wakefulness, the associated EEG is almost always much faster in frequency and often shows the more rhythmic structure seen in alpha rhythms.[11] Very slow EEG frequencies like delta and theta do occur in tandem with much higher frequencies during wakefulness and then may have obvious rhythmic structure, as to be reviewed in this chapter.

The effects that are shown in Fig. 6.1 corresponding to levels of arousal in humans are best seen in more posterior scalp electrodes that overlay occipital, parietal and temporal lobes shown in Fig. 6.2. These underlying brain areas reflect the visual, auditory and somatosensory primary cortex areas, along with secondary and association cortex areas that are sensitive to higher levels of sensory processing and mixtures of these modalities. For this reason, more posterior scalp electrodes are especially sensitive to the influx of sensory information as happens most prominently during the excited state shown in Fig. 6.1. Because the posterior cortex is more closely tied to sensory inputs, the alpha rhythm in humans is most readily observed during resting, eyes-closed wakefulness in these posterior areas. Differentiation of underlying rhythms is also seen during relaxed wakefulness, as the alpha rhythm is much less obvious in more frontal regions where faster rhythms would be more apparent.

The greatest differentiation of underlying rhythms is reflected in the excited EEG state that is associated with higher levels of arousal in an awake

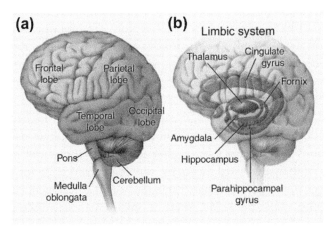

Figure 6.2 Principal areas of the human brain seen from (a) a lateral view and (b) a medial view that shows the Limbic System. Notice that structures within the medial temporal lobe including Hippocampus and Parahippocampal Gyrus are depicted on the right. *Source: Public Domain License Granted at Wikimedia.org.*[12]

or dreaming brain. This excited state was originally referred to as a desynchronized EEG because original pen recordings low pass filtered the higher frequency synchronized gamma oscillations. However, very periodic higher frequency signals which are often in the gamma frequency range (>30 Hz) are now observed in today's digital recordings measured from the scalp. Whereas the slower EEG rhythms are thought to be associated with oscillating neural synchrony involving greater spatial regions, faster EEG signals reflect much less global synchrony and much more isolated pockets of oscillating synchrony which may even be out-of-phase locally so that the voltage trace in the scalp recording shows very low amplitude.[13] Thus, very fast gamma oscillating synchrony still may be difficult to detect in many cases in today's scalp recording and only may be seen in invasively recorded EEG as is done to monitor epileptic patients. Long-range synchrony of periodic gamma activity, such as between frontal and posterior cortical regions, is the best marker of consciousness in animals and humans undergoing general anesthesia. In this case, it subsides during loss of consciousness and only returns during wakefulness.[14]

2. NEURAL SYNCHRONY, PARSIMONY, AND GRANDMOTHER CELLS

At some level, neural synchrony must be present in how the brain represents information. For example, when synchronously activated presynaptic neurons converge and cause a postsynaptic neuron to fire, the postsynaptic neuron's firing binds information together from the input neurons. In the sense where it is defined by synchronous presynaptic inputs that lead to postsynaptic convergence and firing, neural synchrony is similar to the concept of hierarchical convergence advanced by Nobel Prize winners David Hubel and Thorsten Wiesel. Based upon their neural recording data, Hubel and Wiesel proposed that visual cortical neurons at higher levels of processing simply sum the inputs from lower levels to create ever sparser representations that express ever higher levels of sensory and perceptual constructs. In their proposed hierarchy, simple neurons code for basic features such as lines of certain spatial frequency and orientations in specific regions of the visual field. Yet, there is still tremendous redundancy in simple neurons' response in visual cortex, as many neurons respond to similar features.[15] More complex cells code over a larger spatial region and some respond to movement; their firing activity would appear to result from the summation of lower level simple cell inputs.[16] The effect of such processing is that fewer total units may need to be active to code representations at the higher levels, so parsimony is a natural outcome.

Yet, if this idea of ever greater sparsity at higher levels of representation were taken to the extreme, the activity of one single neuron would code for the image of your grandmother. Such extreme sparsity appears to be unlikely though, as an ensemble of synchronously activated or inactivated neurons that operate as a neural synchrony unit that may be called the neural ensemble would seem instead to represent information at each stage. In fact, there is now evidence for an ensemble level neural code which discriminates well between different orientations of visual sine wave gratings using logistic regression with the inputs simply being the relative response of neurons in the primary visual cortex.[17] While one specific grandmother cell that uniquely codes for the image of a grandmother has never been found, much greater sparsity seems to exist with higher levels of representation. That is, ever greater selectivity of representations is found in ever sparser populations of neurons in higher cognitive processing. Thus, in later perceptual processing in inferior temporal cortex, there are a relatively sparse group of neurons that increase firing more so to the face of grandmother, but this response may vary based upon the orientation of the grandmother's face and these cells also may respond somewhat to other faces.[18] At even higher levels of representation in the brain's medial temporal lobe explicit memory system involving the hippocampus and entorhinal cortex, more specific and even sparser coding has been found. In the medial temporal lobe, cells have been found to respond specifically to the face of actress Jennifer Aniston in various orientations, her name, and her famous costar—Brad Pitt, but not to another famous actress—Julia Roberts. In the same study, another medial temporal lobe ensemble of cells showed similar specific firing to another Hollywood actress—Halle Berry.[19]

3. GESTALT *PRAGNANZ* AND OSCILLATING NEURAL SYNCHRONY

This idea of synchronous data-driven hierarchical convergence as a mechanism of cognitive processing such as perceptual representation or explicit memory reconstruction would at first glance seem reasonable as long as it is accomplished in a sparse population of neurons in a neural ensemble rather than through single neurons. However, the Gestalt psychologists in the early twentieth century pointed out that our perception of the world cannot simply be a data-driven process and must instead also result from a higher level binding process where the unitary whole is not simply the sum of the parts. The most basic Gestalt law concerns *Pragnanz*, which in German means "good figure". This idea is that we tend to see the world always in the simplest and more organized way possible and this may in fact distort physical reality. This Gestalt principle implies that our conscious brains always try to force parsimony and good organization onto representations to result in relatively few meaningful high-level perceptual components, as compared to more complex nonsensical representations.

The Kanizsa Square depicted in Fig. 6.3 is a prime example of what the Gestalt psychologists meant in their Law of *Pragnanz*. The white square is actually a perceptual illusion and a distortion of physical reality, as there are not the correct elementary features here to code for a white square. That is, the sides of the square are missing. For this reason, data-driven machine learning may fail completely to represent what humans perceive. A parsimony principle could help to explain human perception in this case. There are five regularly shaped objects in our perception: a bright-white square in the foreground, and four circles in the background. However, there are actually many separate irregular and fuzzy objects in the physical picture, as there are four sets of a large number of fuzzy concentric rings that are missing one-fourth of their structure in right angles. But, the brain perceives a far simpler, more meaningful, and more organized image, which is a white

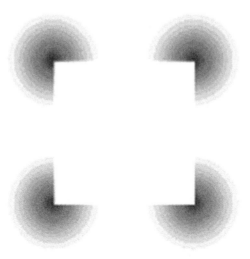

Figure 6.3 The Kanizsa Square.[20]

square in front of four fuzzy circular objects. Thus, the brain constructs a more parsimonious and meaningful perceptual interpretation that is not the actual physical reality. The Gestalt psychologists proposed that this Law of *Pragnanz* is a fundamental process in everything that we perceive. Max Wertheimer, one of the three prominent Gestalt theorists early in the twentieth century, used the following example to illustrate this *Pragnanz* principle:

I stand at the window and see a house, trees, sky. Theoretically I might say there were 327 brightnesses and nuances of color. Do I *have* "327"? No. I have sky, house, and trees.[21]

The principle of parsimony along with hierarchical integration could explain the Kanizsa Square illusion, along with why the brain sees a sky, house and trees. This would have to assume that a relatively sparse ensemble of neurons in the brain is somehow able to represent Kanizsa Square or the sky, house and tree through mutual synchronized oscillations that ensure that the perception is composed of meaningful features. These parsimonious conscious representations could originate with highly redundant feed-forward signals that code lower level unconscious sensory input, such as simple cell activity in the primary visual cortex. As processing moves to secondary visual cortex and association areas in the brain, the redundancy would lessen, as the brain would gradually increase its specificity by selecting the most meaningful features through feedback that removes the most redundant features. Ultimately, this neural ensemble would code for a complex perception as reflected by momentary highly periodic oscillating local field standing wave signals that allow mutual synchrony across multiple regions including primary visual cortex, secondary visual cortex, and association cortex. This oscillating neural synchrony ensemble represents a unitary perception built with higher level meaningful features that have prior memory weights including a white square in the foreground and concentric dark fuzzy circles in the background.

This is one idea about how perception could work. Yet, it is not really the Hubel and Wiesel idea of feed-forward hierarchical integration because it requires feedback to select the final important features and explain their mutually synchronized oscillatory representation, and because it superimposes a principle of parsimony along with meaningful prior memory weights in the final selected features that are the basis of the perception. On this point, the insightful book by György Buzsáki entitled *Rhythms of the Brain* argues that a most basic failing of Hubel and Wiesel's hierarchical integration model of perceptual binding is that it fails to consider the extensive oscillating synchrony feedback signals that are obviously involved in much of the brain's conscious processing.

In fact, such a sparse oscillating neural synchrony mechanism has been proposed by Wolf Singer to be a very basic and general conscious organizing principle that he calls "binding by synchrony".[22,23] This brain process would bind elementary attributes like the right angles and sides of a square to create the experience of a bright-white square in the Kanizsa illusion. Singer has pointed out that Gestalt ideas about perception played a large role in his thinking about perceptual binding.[24] Singer's ideas are based upon findings originally reported in the 1980s that highly periodic local field oscillations in the gamma frequency range were synchronized with zero phase lag across spatially disparate primary and secondary visual cortical regions in cats who were viewing coherent stimuli such as a single moving long light bar.[25] Such perfect oscillating synchrony did not occur during control conditions when two light bars were moved in opposite directions or when two light bars were moved in the same direction, even though these light bars were also exciting the identical visual cortex neurons as the single moving light bar. What was remarkable about these observations is that the gamma frequency oscillations were so periodic and that the synchrony between such very periodic gamma oscillations occurred across spatially distinct regions specifically in the one condition when the conscious experience of seeing one unitary stimulus with coherent and continuous motion should occur. This work suggested that such perfect oscillating synchrony across spatially distinct brain regions represented global properties of the stimulus such as coherent continuous motions that were perceived in a unified conscious experience. This work also suggested that this long-range oscillating synchrony in the gamma frequency range across spatially distinct cortical regions simply could not be explained by hierarchical convergence.

A demonstration of how RELR may model a neural ensemble in the representation of a square which may have missing sides like in the Kanizsa Square illusion is the following. First, an RELR model trains on input data that present a square on half the learning trials and a right triangle that is either oriented with the 90° angle at the bottom left or top right on the other half of the trials. Figure 6.4 shows these square and triangle training stimuli. The input signals to the neural ensemble here could be the ensemble of neurons that code the important visual features of squares and triangles like the presence of 90° or 45° angles, the presence of sides of the figures, along with the important auditory feature "square" or "not Square" which

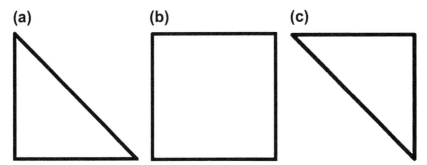

Figure 6.4 Stimuli used to train RELR's square perception. The stimuli in (a) and (c) represent the right triangles used, and (b) represents the square that was used.

is the "verbalization" feature driven by an external instructor or by reading. Note that this verbalization signal is simply a neural representation of the subjective auditory verbalization of the word "square" or "triangle" as might be expected to occur when performing a task to discriminate between squares and triangles. The output signals from this neural ensemble would be measured in population of neurons that produce binary signals in synchrony with these input features, although the probability of the binary output responses in this ensemble would be a closer representation of the local field potentials. This is because the scalp EEG and other measures including intracranial measures of the local field potentials are not sensitive to actual binary spiking responses and instead represent the probability of binary responses in local neural ensembles.[26] Together this neural ensemble that encompasses both input features and output responses may represent the perceived and labeled figure whether it is a "square" or "triangle". The behavior of a single modal neuron within this ensemble is simulated in the toy model training conditions illustrated in Tables 6.1 and 6.2.

Table 6.1 shows the training observations for this bursting neuron with 10 training observations equally split between triangle and square presentations, along with the input visual features and auditory verbalization features, and the binary response output. The binary response is purposefully designed here to not completely depend upon the input features by randomly setting two binary output signals to be incorrect. This is meant to model the possibility of local field potentials having a causal effect on binary spiking responses that is independent from the effects of input features based upon evidence reviewed in the last chapter.[27] The top panel of Table 6.2 shows the regression weights for this model neuron across the input features after the training condition, the retraining condition with the identical

Table 6.1 Initial training episode for RELR square perception. Each row corresponds to one training observation where either a Square or Triangle is presented with the physical features being indicated as "Present" 1 (Y) or "Not Present" 0 (Y) and the Verbalization feature indicates "Square" 1 (S) or "No Square" 0 (T). The output is a binary response that represents either a neural spike (1) or no-spike (0) indicating Square Perception (1) or No Square Perception (0).

TopRt90°	BottomRt90°	TopLt90°	BottomLt90°	TopLt45°	BottomRt45°	Bottom line	Left line	Hypotenuse	Top line	Right line	Verbalization	Binary response
1(Y)	0(N)	0(N)	0(N)	1(Y)	1(Y)	0(N)	0(N)	1(Y)	1(Y)	1(Y)	0(T)	0
1(Y)	1(Y)	1(Y)	1(Y)	0(N)	0(N)	1(Y)	1(Y)	0(N)	1(Y)	1(Y)	1(S)	1
0(N)	0(N)	0(N)	1(Y)	1(Y)	1(Y)	1(Y)	1(Y)	1(Y)	0(N)	0(N)	0(T)	0
1(Y)	1(Y)	1(Y)	1(Y)	0(N)	0(N)	1(Y)	1(Y)	0(N)	1(Y)	1(Y)	1(S)	1
0(N)	0(N)	0(N)	1(Y)	1(Y)	1(Y)	1(Y)	1(Y)	1(Y)	0(N)	0(N)	0(T)	0
1(Y)	1(Y)	1(Y)	1(Y)	0(N)	0(N)	1(Y)	1(Y)	0(N)	1(Y)	1(Y)	1(S)	1
1(Y)	0(N)	0(N)	0(N)	1(Y)	1(Y)	0(N)	0(N)	1(Y)	1(Y)	1(Y)	0(T)	0
0(N)	0(N)	0(N)	1(Y)	1(Y)	1(Y)	1(Y)	1(Y)	1(Y)	0(N)	0(N)	0(T)	1
1(Y)	1(Y)	1(Y)	1(Y)	0(N)	0(N)	1(Y)	1(Y)	0(N)	1(Y)	1(Y)	1(S)	1
1(Y)	1(Y)	1(Y)	1(Y)	0(N)	0(N)	1(Y)	1(Y)	0(N)	1(Y)	1(Y)	1(S)	0

Table 6.2 Initial training and update weights and output probabilities corresponding to square perception training conditions. In top panel regression coefficient weights in RELR model are shown as initial training weights, update weights (no missing data) with identical training sample as initial training, and update weights (missing sides and verbalization) when features corresponding to the square sides (top line, bottom line, right line, and left line) and the verbalization have missing values but the training sample is otherwise identical to the initial training. Bottom panel shows output probabilities for binary response for all 10 observations in these conditions.

Feature	Initial training weights	Update weights (no missing data)	Update weights (missing sides and verbalization)
TopRt90°	0.060	0.000	0.002
BottomRt90°	0.203	0.000	0.013
TopLt90°	0.203	0.000	0.013
BottomLt90°	0.159	0.002	0.014
TopLt45°	−0.203	0.000	−0.013
BottomRt45°	−0.203	0.000	−0.013
Bottom line	0.159	0.002	0.000
Left line	0.159	0.002	0.000
Hypotenuse	−0.203	0.000	−0.013
Top line	0.060	0.000	0.000
Right line	0.060	0.000	0.000
Verbalization	0.203	0.000	0.000

Binary response	Probability w/initial training	Probability w/retraining (no missing data)	Probability w/retraining (missing sides and verbalization)
1	0.125	0.124	0.212
2	0.817	0.817	0.758
3	0.233	0.234	0.263
4	0.817	0.817	0.758
5	0.233	0.234	0.263
6	0.817	0.817	0.758
7	0.125	0.124	0.212
8	0.233	0.234	0.263
9	0.817	0.817	0.758
10	0.817	0.817	0.758

features, and then after the retraining condition with missing sides and verbalization features. Both retraining models are updated with RELR's sequential online learning method based upon minimal K–L divergence/ maximum likelihood as previously outlined in Chapter 3 where the posterior weights from the initial training episode serve as the prior weights for retraining/updating conditions. The bottom panel of Table 6.2 shows the binary output probabilities corresponding to these 10 observations in the three training conditions. Notice that because RELR is easily able to handle missing values, it also easily perceives the squares instead of triangles when sides are missing as input features. The update weights in the "missing sides and verbalization" condition are all zero because RELR fills missing values with zeroes in its standardized features, and the update weights in the other features of the square are minimally changed. As reviewed in Appendix A4, the new posterior weights after the update/retraining condition is the sum of the prior and update weights. So, it would take many such retraining conditions with missing sides and verbalization before this RELR model would substantially change the relative weights in the posterior weights of the features that are present after retraining with missing data. These training and retraining episodes require the assumption that all training observations within an episode are independent in the sense that one observation cannot cause another observation; such an assumption might be reasonable, for example, in the case of a neuron that has rapidly bursting output signals. If this assumption is not met, then both training and retraining would need to occur at the pace of one observation at a time.

Perceptual binding by synchronous oscillating feedback gamma frequency loops would require a large number of different synchronized and oscillating neural ensembles to represent every possible perception, but that is possible. In fact, a scaled down model has been built by Werning and Maye.[28] When neural ensembles oscillate in synchrony in this model, these ensembles represent a common object; when these oscillations are desynchronized, they do not. Though simultaneously activated, the different oscillating components are almost uncorrelated across time and there is one larger more holistic oscillating component that contains evidence of all other components. For example, one oscillating component may represent head and arms, another represents body and legs, and the holistic component contains evidence of an entire body. The Werning and Maye model is unlike the artificial neural network models of a generation ago, as this model is not at the single neural model but instead represents the activation of an ensemble of neurons in local synchronous oscillating circuits within visual cortex. Whether this model is ultimately accurate, it is useful in showing how approximately orthogonal oscillations can be synchronized to generate a cognitive representation that seems to obey the Gestalt principle that the whole is greater than the sum of the parts.

Long-range oscillating gamma frequency synchronization between neural populations in very distinct regions of the brain such as across hemispheres has also been observed during visual perception binding.[29] There is however controversy as to the extent to which long-range synchrony could explain the cognitive binding of elementary attributes into larger attributes. Some studies that have looked for long-range synchrony in spike response recordings during visual perception have not found it.[30,31] Singer has recently commented that all evidence is that binding by synchrony effects are much easier to see in EEG field recordings than in single neuron spike-response recordings.[32] This makes sense because oscillating and synchronous activity is also much easier to see in EEG field recordings than in spike responses, as spike responses are much noisier. Note that traveling waves of synchronized EEG field correlations across spatially distinct cortical regions actually have been observed during visual processing, so strict long-range synchrony in perfectly standing waves always may not be involved in binding.[33] Instead, some models have allowed for slight phase lags between spatially distinct regions consistent with traveling waves that do not always combine to create perfectly stable standing waves.[34,35] This is similar to how some parts of a musical orchestra often echo other parts with slight delay, but the effect is clearly experienced as a unified musical theme.

Singer has suggested that oscillating neural synchrony functions to provide shorter term and more flexible synchrony than afforded by fixed anatomical connections between neurons. While Singer agrees that much of oscillating gamma frequency synchrony could occur through fixed short-range anatomical connections such as interneurons in cerebral cortex, he argues that the critical binding mechanism of longer range synchrony must rely upon the periodicity of oscillations. This is because the longest range neural synchrony effects occur too rapidly in the higher gamma frequency range (>30 Hz) to be determined by fixed neuronal connections.[36] As an example of how this periodicity may be used to determine long-range synchrony, Buzsáki has argued that periodic oscillations would be a very inexpensive means for synchronous parallel processing between two distant neural regions to be accomplished. If two brain regions have coordinated oscillating periodicities, then they are naturally synchronized and do not have to communicate to one another in real time, but instead could rely upon intermittent and slower long-range communications to maintain synchrony. This mechanism of long-range oscillating synchrony achieved through intermittent communications also might explain slower brain rhythms than in the gamma frequency range like those in the theta (4–8 Hz) and alpha (8–12 Hz) range which are known to depend upon fixed corti-cohippocampal or corticothalamic neural connections.

The only way that coordinated repetitive temporal patterns in oscillating neural synchrony could occur with two or more independent oscillators is if there is approximately an orthogonal relationship between the different oscillators which would manifest as small whole number frequency ratios like 1:1, 2:1, 3:2, or 4:1. In fact, there is abundant evidence for such simple small whole number ratio structure in oscillating neural synchrony as will be reviewed shortly. In perceptual processing, these highly periodic and harmonic oscillating neural synchronies could function to organize the large amount of disconnected sensory information into a few meaningful conscious chunks of perceptual information like sky, trees and house. In higher level explicit working memory processing, oscillating neural synchrony between different elements in the neural ensemble could explain how the mental representation of Brad Pitt can occur simultaneously or in close temporal succession with the image of Jennifer Aniston, as perhaps because there is explicit recollection that one of the features of Jennifer Aniston is that she was married to Brad Pitt or that she was in the same movie as him.

In building up parsimonious explicit cognitive representations from simple oscillating feature elements that exhibit small whole number ratio structure, this music-like neural ensemble orchestra may be an embodiment of the Law of *Pragnanz* that requires that conscious representations are always as simple and meaningful as possible. Explicit RELR's parsimonious feature selection will tend to discard highly multicollinear feature elements and instead returns features that are as uncorrelated or orthogonal as is possible while still fitting the data. This was exemplified back in Fig. 3.4(b) where the final Explicit RELR model discarded a feature that was highly correlated with the final selected feature. So, Explicit RELR also will tend to return as simple and meaningful a representation as is possible while still fitting the data in accord with *Pragnanz*, especially because its selections also can be determined by meaningful prior weight parameters in online sequential learning. Yet, a big difference between Explicit RELR's treatment of multicollinear features compared to standard predictive modeling methods like principal component analysis (PCA) is that RELR does not force orthogonal features as predefined features. In fact, more complicated Explicit RELR models often will contain highly multicollinear features if needed to fit the data. In contrast, Implicit RELR will almost always contain highly multicollinear features if they are predictive because its objective is not determined by parsimony principles. The problem with forcing orthogonal features is that the solution might be entirely an artifact of an assumption of orthogonal features. For example, an algorithm like Fourier analysis that forces orthogonal solutions will find orthogonal solutions with frequency components in perfect whole number ratios even in white noise observations. On the other hand, when Explicit RELR finds simple structure that is built from features that are relatively uncorrelated, it is much more likely to be a stable structure that would be similar if the model had been trained with independent data.

4. RELR AND SPIKE-TIMING-DEPENDENT PLASTICITY

Spike-timing-dependent plasticity effects are observed quite widely in neural learning. These effects are evidence for the Hebbian learning principle introduced in the last chapter.[37] The basic effect is if a presynaptic neuron spikes within a brief window of time prior to when a postsynaptic neuron spikes, this will have a higher probability of causing long-term potentiation (LTP), which is a long-term positive weight change in an input learning weight. On the other hand, if the presynaptic neuron spike occurs in a brief window of time after the postsynaptic neuron spikes, synaptic long-term depression (LDP) and a more negative learning weight change is likely. Spike-timing-dependent processes were originally observed in excitatory synapses; yet these processes are now known to exist in inhibitory synapses.[38]

Temporal coincidence involving the relative timing of pre- and post-synaptic spikes is involved in spike-timing-dependent learning effects. The rapid learning of new connections through temporal coincidence of pre- and postsynaptic activation would allow for flexible rerouting of information that does not depend upon prior fixed anatomical connections. The resulting spike-timing plasticity effects could lead to LTP and LDP and have the effect of boosting short-term activity in those synapses that are most active and important for memory consolidation. Consistent with these results is the observation that human memory is strongest when there is a strong coupling between local theta oscillations and neuronal spiking in medial temporal lobe.[39]

Abundant empirical evidence indicates that neural learning weight changes are related to tight temporal correlations between pre- and post-synaptic spikes. Yet, the relative positivity or negativity of graded potentials in dendrites appears to be the critical effect rather than the timing of the spikes in presynaptic neurons relative to spike in postsynaptic neurons.[40,41]

When spikes in postsynaptic neurons occur during greater positivity in their dendritically measured potentials, LTP appears to be more likely in these same dendritic synapses. In contrast, when spikes in postsynaptic neurons occur during greater negativity in their dendritic measured potentials, LDP appears to be more likely. The precise mechanisms that cause these LTP and LDP weight changes are still being worked out and may relate to a back propagation mechanism in dendrites.[42]

Like this neural spike-timing-plasticity mechanism, RELR's learning is based upon the tight correlation between input features and binary signals. Those features which are considered as viable candidates tend to have the highest magnitude correlations in either a positive or a negative direction. Likewise, RELR's maximum entropy logistic regression learning mechanism is based upon a cross-product sum across binary responses and coincident independent variable feature values (Appendix A1, Eqns A1–A5). This mechanism also updates prior memory weights through minimal $K–L$ divergence learning (Appendix A4). This cross-product sum is computed for each independent variable feature to form a set of basic constraints in the maximum likelihood logistic regression optimization. This RELR learning mechanism is consistent with spike-timing-dependent plasticity that is caused by greater relative positivity or negativity in the features. In a RELR neural mechanism, postsynaptic spiking responses that occur during greater feature positivity would be more likely to give greater positive weight for a given effect, whereas greater negative weight would occur as a result of spiking that occurs during greater feature negativity.

Thus, RELR's logistic regression learning may have reasonable similarity to the spike-timing-dependent plasticity in real neural learning at least in an aggregate probabilistic sense. The sum of the cross-products between a graded potential input effect and the axonal hillock binary spike signal across a training episode time interval would be greater for more synchronous input effects. This cross-product sum is identical to the cross-product sum that is the basic constraint that determines weight learning in RELR. This is depicted in Fig. 6.5, which shows that the advantage of synchronous graded potential inputs would be that they would produce more positivity in tandem. Likewise, if they occurred during negative phases of the local field potentials, they would produce more negativity in tandem. Therefore, as also happens in neural spike-timing-learning mechanisms, the RELR learning mechanism is general enough to handle spike-timing-dependent processes that are excitatory, inhibitory, or that change direction during the course of the development.[43] This is because RELR's learning is entirely

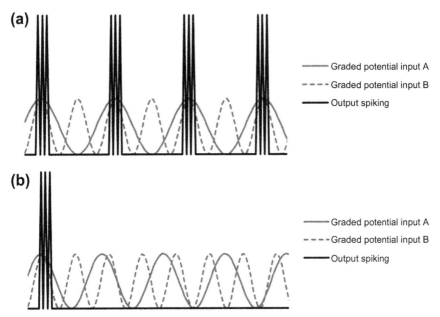

Figure 6.5 (a) Separate synchronous graded potential inputs are shown in relationship to binary output spiking. (b) Separate asynchronous graded potential inputs are shown in relationship to binary spiking.

data-driven and caused by correlations between independent variable features and binary responses, which may change direction over time.

5. ATTENTION AND NEURAL SYNCHRONY

The sparse oscillating neural representations of conscious human perception, memory and thinking are ultimately a function of what is amplified by the attention system. Attention can be described as a form of filtering where certain signals are allowed to pass through to conscious processing and others are prevented. A more aesthetically pleasing metaphor that is sometimes used by psychologists is that attention is the conductor in the orchestra of explicit cognitive experience, as it allows explicit cognition to exert control over implicit cognitive information and focus on what is truly important. In essence, this conductor is constantly directing all instruments in the brain's neural ensemble orchestra, but also extemporaneously composing its own oscillating synchronies, periodicities, and rhythms.

Psychologists can measure attention processing with reaction time and signal detection performance. During the excited EEG state depicted in

Fig. 6.1, attention to external stimuli is heightened as measured by faster reaction times and better signal detection than during the relaxed and wakeful times when the alpha rhythm predominates. Attention also fluctuates as a function of the phase of the resting posterior alpha rhythm as measured by auditory and visual performance in humans. The resting alpha rhythm is thought to reflect extensive synchronous feedback oscillatory signaling between the thalamus and the cortex. As shown in Fig. 6.2, the thalamus sits in the center of the brain as a passageway between sensory organs and the cerebral cortex. Different thalamic nuclei serve as a major relay station for sensory information on the way toward the primary cortex for the different modalities. Yet, the thalamus is not simply a passive responder to incoming sensory inputs. Instead, the oscillating neural synchrony as measured in the alpha rhythm would appear to gate input information from the physical environment. This is essentially a sampling mechanism that does not continuously monitor the external world equally at all times while we direct attention inward, but instead samples the external environment with a sensitivity to incoming information that varies directly with the phase and frequency of the alpha rhythm at roughly 8–12 Hz.[44] Even during the excited EEG state, the evidence is that the phase of the EEG still determines attention, as detection performance has been observed to vary with the phase of a 7 Hz theta rhythm in this case.[45] Thus, oscillating neural synchrony may be able to serve as an attention shutter rhythm that opens and closes to determine the sensory information that is processed by higher cognitive functions in the brain.

This attention mechanism also can be studied at a lower level like in a single neuron or a small ensemble of neurons. This research suggests that attention is regulated by neural synchrony. The same neural synchrony principles that may apply to a larger ensemble of neurons seem to apply to the individual neuron. In particular, input neurons that exert synchronized effects on an output neuron appear more influential and have an amplified effect on the output; a similar mechanism can be assumed to produce attention in a neural ensemble representing groups of neurons.[46] These results would be predicted at a neuronal level because synchronized spiking presynaptic neurons that produce synchronized graded potential effects in the postsynaptic output neuron would be expected to exert much stronger effects on the output neuron's axonal spiking compared to asynchronous inputs. Thus, greater synchrony of excitatory neural inputs should be associated with enhanced spiking compared to asynchronous inputs as simulated in Fig. 6.6. In neural computation, binary spiking and input

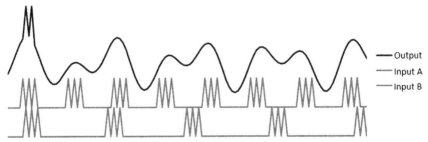

Figure 6.6 At the top is a simple simulation of the compound output signal at the soma that would arise from excitatory synchronous neural spiking inputs. At the bottom is the same such output signal that arises from excitatory asynchronous neural spiking inputs. This is designed to be a simple integrate-and-fire response with an all-or-none spiking output. Notice that much greater spiking can occur in response to two excitatory synchronous inputs compared to two excitatory asynchronous inputs. The 2:1 frequency ratio between the synchronous inputs is not accidental in the top panel, as oscillating neural synchrony would arise through simple whole number frequency ratios in the firing response of the neurons that form the inputs.

feature signals that are perfectly in phase or perfectly out of phase would imply strong positive or strong negative correlation between binary spiking and features. On the other hand, random phase relationships between input features and binary spiking would be associated with much smaller magnitude correlations and corresponding lower t-value magnitudes. So, this attention mechanism may be similar to the RELR feature reduction based upon t-value magnitudes which serves to focus on the most important candidate features for later processing. These attention neural synchrony computational effects are shown in Fig. 6.6.

Another well-studied attention phenomenon is that the brain habituates very rapidly to monotonous stimuli whether it is a monotonous tone, a monotonous spoken word, or monotonous visual imagery in driving along a highway. This indicates that the attention mechanism greatly prefers novel stimuli and will ignore monotonous stimuli whenever possible. When only monotonous sensory inputs are available when driving on a monotonous highway, people often become drowsy and fall asleep. A similar attention preference for novel or rare events is seen in RELR's preference to balance rare versus frequent target outcome events in training samples, as a sample composed of all outcomes of one value would be meaningless.

Whether attention is considered at the neural ensemble or individual neuron level, this still may be considered to be a feature reduction process that also may sample the learning episode to prefer more novel observations. Attention seems to work something like how caching works in hard disk processing. It breaks incoming data into learning episodes that are the basic data epochs, and it is those episodes that are processed in working memory. To some extent, it is driven by external events, as when a relatively rare event in the presence of monotonous stimuli awakens attention. Yet, the brain's attention is also inextricably connected with explicit prior memory as in the square perception example. In addition, the brain's attention is also under the control of the central executive function in working memory, as in the top-down processing that originates in the frontal eye fields of pre-frontal cortex and gradually restrict visual inputs in structures that eventually include the primary visual cortex.[47] Attention of cached samples of incoming data allows the brain to engage in sequential online learning while memory weights are constantly updated using small samples of new ob-servations in ways that could be similar to RELRs sequential online learning. The human brain clearly does not perceive information in real time. This is exemplified by film which is built from 20 discrete frames per second, but where continuous motion images are perceived. Thus, real-time learning might not be necessary to simulate the brain. Instead, rapid cached sequential working memory updates might be all that is necessary.

The important point is that the brain does not seem to perform batch sample learning across huge numbers of observations which is a lesson that might be of value to current machine learning notions about how Big Data might be used to yield smarter artificial intelligence. Instead of using enormous Big Data observation samples across lengthy historical periods, the brain seems to sample brief episodes in time that are the basis of working memory. The memory of these episodes are likely to become episodic longer term memories if rehearsed or processed more deeply, although the consolidation of recent explicit memories into substantially longer term explicit memories also may depend upon effects observed during sleep.[48]

6. METRICAL RHYTHM IN OSCILLATING NEURAL SYNCHRONY

The lower frequency EEG rhythms that are typically prominent during human wakefulness are <30 Hz and include theta, alpha, beta, and lower frequency gamma rhythms, although even very low-frequency rhythms in the delta frequency range are observed at times. These low-frequency brain signals often seem to reflect the duration and rhythm of separate conscious processing events much like the duration and rhythm of musical notes. In fact, separate low-frequency brain signals can have small whole number duration ratios which are also akin to the rhythmic structure in many forms of music. An example is the 4:1 ratio that is observed between the duration of quarter and sixteenth notes in metrical rhythm in western music.

A striking example of this low-frequency rhythmic structure in the EEG is from a study that concerned the activity of the hippocampus in epileptic patients by Axmacher and associates.[49] Patients were engaged in a working memory task that is a modified form of the well-known Sternberg memory task. Novel faces were first serially presented to the patients with one, two or four items in a to-be-remembered set. Next, there was a 3.5 s maintenance period when these items were to be rehearsed. Finally, a probe face was presented and patients had to indicate whether or not the probe face was in the preceding set of to-be-remembered faces. During the maintenance phase of this task, there was significant phase and frequency coupling of rhythms in the theta range with a peak spectral power at 7 Hz and the beta/gamma range with a peak spectral gamma power at 28 Hz. That is, the faster gamma rhythm was amplitude modulated by and phase locked to the positive phase of the slower theta rhythm. Most interestingly, the rhythmic structure varied with the number of items to-be-remembered in this task, as one item related to one gamma cycle per theta cycle, two items related to two gamma cycles per theta cycle, and four items related to four gamma cycles per theta cycle. Also very notable, the frequency ratios of gamma to theta rhythms remained locked at 4:1 across these memory conditions. Although there was a shift toward slower theta rhythms with increasing memory load, the gamma signals also trended toward slower rhythms as memory load increased from one to four items. The simultaneous slowing in both theta and gamma rhythms with increasing memory load is why the 4:1 ratio in frequency remained in effect across the memory load conditions. This is a very rapid extreme tempo during this challenging memory task, but it is close to the upper range of what might observed in American Jazz music.

Based upon other data involving this coupling between theta and gamma rhythms during working memory, Jensen and Lisman had proposed earlier that the number of gamma cycles that are coupled to the theta rhythm should reflect the number of items in a working memory maintenance rehearsal process. That is, the processing of each to-be-remembered item would be reflected by each different gamma cycle in a gamma rhythm that is itself locked to the theta rhythm. Thus, if a person rehearses a seven–digit phone number, there should be seven gamma cycles per theta cycle and each gamma cycle would represent the timing of the brain's processing of a different digit in the series. Likewise, the rehearsal of four facial images would be associated with four gamma cycles per theta cycle, whereas the rehearsal of one facial image would be associated with only one gamma cycle per theta cycle. The Axmacher et al. findings are interpreted to be in support of this model that every gamma cycle in a sequence reflects the processing of an independent object of consciousness in the underlying neural ensemble. These rehearsed neural ensemble objects are then synchronized to the theta rhythm in the hippocampus during the brain's working memory processing.[50]

However, the Axmacher et al. finding of an invariant small whole number frequency ratio across the memory load conditions does contradict one detail in the earlier Jensen and Lisman model. Jensen and Lisman had proposed that the theta rhythms should slow to accommodate more rehearsed items but the gamma rhythm should not, as gamma was assumed to have a constant frequency in this model so that whole number ratios would not be maintained between these two frequencies. In contrast, this Axmacher et al. finding that the theta and gamma rhythms are organized in invariant whole number frequency ratios suggests that this EEG rhythmic structure is similar to metric rhythmic structures seen, for example, in western music where quarter and sixteenth notes exhibit an approximate constant 4:1 duration ratio across varying tempos. Wavelet analysis is typically in studies like Axmacher et al. because it is sensitive to such temporal organization and variation in rhythms.

Another interesting finding in the Axmacher et al. study was that there was significant coupling between very low-frequency delta (1–4 Hz) and the beta rhythm (14–20 Hz) although there was no increase in this coupling during the Sternberg maintenance rehearsal phase compared to baseline. Thus, other faster rhythms may be organized relative to other slower beat rhythms in analogy with metrical rhythms in music, so this metrical rhythm structure would not be expected to be unique to gamma and theta synchrony. Indeed, this metrical rhythm organization may be present across diverse cognitive operations in the brain, and also not be unique to the brain's working memory rehearsal.

The Axmacher et al. findings are based upon within-subject comparisons of hippocampus recordings during working memory rehearsal in humans. Kaminski, Brzezicka and Wróbel[51] also studied the association between human working memory and the theta/gamma ratio, but they used scalp recordings and were interested in between-subject comparisons. Kaminski et al. report that individual differences in working memory capacity as measured in a modified Wechsler digit span task are directly correlated to gamma/theta ratio. This digit span test assesses one's skill in successfully reproducing serially presented digits such as a phone number, so it is also a measure of working memory rehearsal ability. Lisman and Idiart[52] (1995) had originally proposed that the gamma/theta ratio determines the capacity of human working memory; Kaminski et al. interpreted their finding as supporting the Lisman and Idiart hypothesis. Across Kaminski et al.'s 17 subjects, this digit span capacity varied between four and eight and there was close to a one-to-one relationship with gamma/theta ratio although it varied more so at the higher ratios. The EEG was scalp recorded from mid-frontal electrodes, and this would be expected to have more noise than the hippocampal measure of Axmacher et al. The higher gamma/theta ratios also would be expected to have more noise, as between-subject comparisons would have greater than noise than within-subject comparisons. Thus, greater noise could explain the lack of a perfect 1:1 relationship between number of memory items and number of gamma cycles coupled to theta in this study. The other possibility is that the best performers chunked two or more digits together, so that they could then rehearse more information more rapidly, and each gamma cycle would then still represent a rehearsal item. Still, overall these findings are supportive of the notion that the gamma/theta coupling reflects the rhythm of working memory processing including rehearsal processing.

Oscillating neural synchrony mechanisms, as reflected in rhythmic coupling between theta (4–8 Hz) and gamma rhythms in the 25–30 Hz range, are proposed to reflect a mechanism that allows the hippocampus and associated medial temporal lobe structures to place shorter term working memory patterns into a longer term store and to retrieve such patterns when needed.[53] When the brain rehearses or maintains items like faces in working memory, it likely does not store all features in rote detail. Instead, it likely stores specific more pronounced features like a large nose. In addition, it may chunk many facial features by storing larger associations to a face like who this person looks like. Thus, it is reasonable to expect that it stores the features and associations that it processes most deeply. In other words, this feedback process that ultimately stores the most important and most meaningful information may be a feature selection storage process akin to an Explicit RELR online sequential learning process that is driven very much by prior meaningful memories. In fact, the learning of new episodic memories has long been known to be enhanced by depth of processing that can utilize the prior meaningfulness of the to-be-remembered items.[54]

A similar rhythmic organization in lower frequency brain signals below roughly 30 Hz has been observed in brain structures not directly involving the medial temporal lobe and during a variety of tasks that can be characterized as involving more than just working memory. For example, Carlqvist and collaborators[55] extracted resting scalp recorded EEG alpha (7.5–12.5 Hz) and beta (15–25 Hz) rhythms from human subjects. They found approximately a 1:2 frequency ratio between the beta and alpha oscillations. The spectral power correlation between these rhythms was strongest in posterior electrodes where both rhythms also had the most power. Thus, the alpha and beta rhythms were coupled much like theta and gamma in invasive hippocampus recordings, but with a different whole number frequency ratio. Nikulin and Brismar also report 1:2 ratios between alpha and beta frequencies in a larger study composed of 176 adults during resting EEG.[56] Palva, Palva and Kaila studied scalp MEG in humans at rest and during mental arithmetic with two or three digits.[57] They found that phase synchrony between alpha and beta or gamma was prominent mainly in the right hemispheric parietal region during the mental arithmetic task where there was a 2:1 or 3:1 frequency ratio of beta (20 Hz) or gamma (30 Hz) to alpha (10 Hz) with phase locking between the two rhythms. The higher frequency ratio was most prominently seen in the more difficult mental arithmetic task involving three digits. This Palva et al. finding again supports the idea that the number of gamma cycles per lower frequency rhythm may reflect the neural ensemble's processing rhythm that represents separate conscious objects computed through oscillating synchrony mechanisms. But in this case, the lower frequency rhythm is the alpha rhythm and the objects are mental arithmetic items and the oscillating synchrony is appearing in the right parietal region.

Roopun[58] and collaborators comment that mental arithmetic effects like observed in Palva et al. might be explained by a memory matching and attention model proposed by Sauseng and associates.[59] Sauseng et al. find that during attention to a visual target, there is enhanced synchrony between a gamma (e.g. a 30 Hz signal) rhythm and a lower frequency rhythm (e.g. an alpha or theta rhythm). They suggest that the gamma and lower frequency alpha or theta rhythms represent bottom-up and top-down processing, respectively. This is interpreted as memory matching between incoming visual information and stored top-down information. Such a mechanism would highlight the function of oscillatory brain activity in the integration of attention and memory processes in working memory. In this mechanism, the increased interregional synchrony of the lower frequency rhythm—theta or alpha—seems to be important in setting the temporal rhythm and duration of this conscious cognitive process, whereas each cycle of the higher frequency beta or gamma rhythms up to about 30 Hz may represent the separate objects of working memory. Small whole number frequency ratios between these various rhythms would be needed to give rise to stable repeating standing wave oscillating synchrony effects in an explicit cognitive mechanism. If bottom-up processing is taken to be the implicit cognitive processing in attention that performs feature reduction and top-down processing is assumed to be the explicit cognitive processing that recalls a given object or feature from memory, then this type of matching mechanism possibly could be a matching mechanism similar to RELR's sequential on-line learning or RELR's outcome score matching process. This is because both these RELR matching processes can be modeled as interplays between implicit and explicit processes. This suggestion is speculative and any resemblance to the brain's sequential and simultaneous matching processes would require empirical data like what may be obtained in the BAM project over the coming years. But, RELR at least allows putative neural ensemble matching mechanisms to be imagined, which may be a starting point for hypothesis testing with better data afforded by the BAM project.

Roopun et al. also comment that other irrational frequency ratios often exist between lower and higher frequency rhythms in the EEG field potentials and thus give rise to nonstationary and nonrepeating signals. If one believes the Buzsáki hypothesis outlined previously that feedback signaling in the brain represents conscious processing whereas feed-forward signaling in the brain represents nonconscious processing, then a simple interpretation is possible. The irrational frequency ratios might reflect underlying feed-forward traveling waves in implicit unconscious cognitive processes that never resonate with more stable explicit conscious periodic standing wave signals. On the other hand, the low-frequency brain rhythms that are related in simple whole number frequency ratios to the higher frequency rhythms would be the same implicit unconscious signals that resonate with conscious object perception and memory events to produce explicit learning. This could occur through a bottom-up/top-down matching mechanism like proposed by Roopun et al. but which is also akin to RELR's model of online sequential learning. Again, empirical tests of these hypothetical neural computation mechanisms might be possible with BAM data in coming years.

7. HIGHER FREQUENCY GAMMA OSCILLATIONS

The available evidence is that lower frequency brain electromagnetic field signals roughly at or below 30 Hz can have music-like metric rhythm structure and that there may be correspondence to the temporal events in conscious cognitive experience. However, low-frequency brain signals would not be able to carry information at a rate that even closely approximates the maximal firing rate of neural impulses of 1000 Hz. On the other hand, higher frequency gamma signals >30 Hz would be able to carry information close to this rate, as invasive recordings suggest that gamma signals range all the way to at least 500 Hz in humans. Since middle C in western music is approximately 264 Hz in its fundamental frequency, higher frequency gamma signals are in the same frequency range as the tones that form the harmonic and melodic structure of music. This high-frequency tonal structure of music is experienced as its most informative aspect, which seems to express conscious emotional and cognitive events in a pure language of oscillating synchrony. This is indicated by Schopenhauer's famous suggestion that the higher tones embodied in the melodic voice represent the "uninterrupted significant connection of one thought from beginning to end" in the "intellectual life and endeavor of man". Thus, a very intriguing question is whether the brain's very high-frequency gamma field potential signals may be rhythmically organized relative to the lower frequency rhythms in a way that is analogous to how higher frequency musical tones are organized with respect to rhythmic lower frequency note patterns. A related question is whether this music-like structure can be observed during informative explicit cognitive processing.

The problem is that higher frequency gamma signals are the hardest to record either invasively or noninvasively due to the low pass filter properties of the brain and scalp tissue. Even with this constraint, there is still a large and growing amount of evidence that the higher frequency gamma signals do actually reflect higher order explicit cognitive processing that would be properly labeled as perception or thought. Recall that the original observation in the 1980s that led to Singer's proposal that gamma oscillations represent perception was based upon highly periodic and synchronized 40 Hz visual cortex gamma oscillations in cats. Similar highly periodic gamma oscillations are observed at higher cognitive levels in humans, but they often have much higher frequencies.

As one example, Gaona and associates show that in subdural and dural electrodes placed over left hemispheric cortical language areas, gamma activity in the range of 80 Hz is more pronounced during speaking a set of words, whereas gamma activity in the range of 288 Hz is more pronounced during hearing the same set of words.[60] Higher frequency gamma signals also may show aspects of music-like organization in that they are rhythmically organized relative to the lower frequency rhythms. For example,

Jacobs and Kahana[61] show that neural representation of individual letter stimuli in humans may be revealed by higher gamma-band invasive EEG activity (up to 128 Hz). Gamma-band activity phase locked to theta increased at specific electrodes and to specific letters. The gamma activation in the occipital cortex seemed specific to the physical properties of the letter because the activation pattern across electrodes was similar for physically similar letters. In other brain regions such as temporal cortex, the physical specificity of gamma increase was not observed as instead the activation seemed to be specific to the abstract symbol. So this is evidence that high-frequency gamma signals are both phase locked to the underlying theta rhythm and are activated during both data-driven physical stimulus processing and semantic processing in cerebral cortex.

Other studies also show that high-frequency gamma signals are phase locked to lower frequency theta or alpha rhythms and this varies with the cognitive task. For example, Canolty and colleagues[62] found that high gamma activity (80–150 Hz) in invasive EEG recordings in neocortical temporal, parietal and frontal regions was phase locked to theta (4–8 Hz) with stronger modulation occurring at higher theta amplitudes. Different tasks elicited different spatial patterns of theta/gamma coupling. Examples of tasks included an auditory Sternberg phoneme working memory task, passive listening to tones or phonemes, auditory-tactile target detection, linguistic target detection, and verb generation. One other important finding was that tasks with overlapping behavioral components exhibited more similar patterns of spatial theta/gamma coupling. Voytek and associates[63] replicated this earlier Canolty et al. finding related to gamma/theta coupling. They also studied gamma/alpha coupling and found that it has somewhat similar properties. The exception was that gamma/theta coupling was more anterior and is more likely to be observed in nonvisual tasks compared to gamma/alpha coupling. These studies add to the impressive body of evidence that high-frequency gamma signals are temporally organized with respect to lower frequency rhythms during conscious cognitive processing.

One notable problem in today's Big Data machine learning is that it is very difficult to implement parallel processing mechanisms compared to serial processing. This is undoubtedly because the rhythm and timing of the separate methodic components need to be well synchronized in much the same way that a conductor needs to make sure that various sections of the symphonic orchestra synchronize together. Perhaps, because of rhythmic structure in parallel processing across time so that components are well harmonized, the brain's constant exchange between explicit and implicit cognitive processing overcomes this problem in ways that would be a good model for machine learning.

So, the brain may use what is recognized as music-like structure in how it organizes its massively parallel signaling into some semblance of conscious cognitive order. These raw field potential signals certainly do not sound like music when they are recorded, as there is a large amount of noise present always. Yet, when filtered, averaged, or recorded locally in a way that amplifies the signal and reduces the noise, rhythmic and periodic properties become clear. These surprising discoveries of oscillating neural synchrony with essentially music-like structure during conscious human cognition seem to harmonize the most grandiose mystical speculations of Pythagoras and Schopenhauer in a leitmotif that is central to modern neuroscience. A constant interplay between consonance and dissonance of underlying explicit and implicit neural ensembles throughout the brain is also consistent with recent evidence that emotion seems to be a property of many simultaneously activated brain regions.[64] Unfortunately, the working memory neural ensemble is not always capable of resolving underlying irrationality and dissonance in ways that even remotely compare to a well-harmonized string ensemble playing a Bach fugue. That is the subject of the next chapter.

CHAPTER 7

Alzheimer's and Mind–Brain Problems

"If I had to live my life again I would have made a rule to read some poetry and listen to some music at least once a week; for perhaps the parts of my brain now atrophied could thus have been kept active through use."
Charles Darwin, Autobiography, 1887.[1]

Contents

In later life, Darwin turned his enormous observational and insight abilities to the problem of cognitive aging. And, he came to the conclusion that his own cognitive deficiencies were related to long habits of overspecialization and disuse. Darwin's explanation for cognitive decline in aging should not be surprising. In fact, this explanation is based upon the same natural selection mechanism that he believed to operate quite generally in all biological systems. In the brain, one may view natural selection as a mechanism that selects those neural ensembles that are most important while discarding those that are not important. Over time, this process should work to build up strong and important neural ensembles that support well-practiced cognitive abilities that would be resistant to age-related decline. As a result, cognitive abilities that are never practiced would be more susceptible to age-related losses in the efficiency of neural ensembles. Thus, while other factors such as genetics and previous brain injury may expedite cognitive dysfunction with advanced age, Darwin's natural selection neuroplasticity mechanism could be an important preventive mechanism.

Today, a nontrivial percentage of elderly can be expected ultimately to get Alzheimer's disease. This is a disease of a lost mind, which causes problematic questions about the relationship between mind and brain.

Calculus of Thought
http://dx.doi.org/10.1016/B978-0-12-410407-5.00007-6

Ultimately, this same mind–brain problem needs to be considered in a realistic assessment of just what is possible with cognitive machines.

1. NEUROPLASTICITY SELECTION IN DEVELOPMENT AND AGING

The brain's natural selection learning mechanism arises through spike-timing-dependent plasticity. All available evidence is that this neuroplasticity is initially caused by neural synchrony mechanisms that lead to long-term synaptic potentiation and depression effects. These potentiation and depression effects are eventually hardwired through changes related to synaptic pruning, synaptic sprouting, and possibly the growth of new neurons in some structures such as the hippocampus. The precise cellular details of spike-timing-dependent neuronal plasticity mechanisms are still being understood, but there is quite general agreement that synaptic pruning must be an important factor earlier in life. In some brain regions such as thalamus and cerebral cortex, the number of synapses may be reduced by as much as 50% or more from infancy until young adulthood.[2–4] This pruning mechanism is thought to occur because synaptic sprouting due to learning during development originally gives overproduced synapses that must be constantly pruned back in a highly selective way. So the mechanism may be a constant alternation of sprouting and pruning that would result in a clear downward trend in total number of synapses between infancy and younger adulthood.[5]

It has been known for many years that a relatively enriched learning environment has the effect of increasing synaptic numbers in the developing brain, so experience-related synaptic sprouting has to be an important part of the neural learning development cycle.[6] Evidence from mammalian learning during motor tasks suggests that dendritic spines sprout together in a spatially clustered manner in motor cortex in a way that seems to encode new memories. Spatially clustered and sprouting dendritic spines are observed to form new spatially clustered synapses over the course of several days. Such spatially clustered synapses are more likely to survive after training compared to nonclustered synapses.[7] The sprouting of spatially clustered synapses is highly task specific in these mouse studies, as those tasks that are repeated are associated with new clusters of synapses, whereas those tasks that are not repeated are less likely to have their associated synaptic clusters survive. Thus, at a cellular level, this alternation of sprouting and pruning seems to depend upon spatio-temporal properties related to synaptic proximity and synchronous repetitive activation that are entirely consistent with oscillating neural synchrony, as newly sprouted synapses that are not repetitively activated with those in close proximity would be pruned.

There are clear developmental trends in the human electroencephalography (EEG) and long-range neural synchrony that parallel the downward trend in total synapses from infancy to adulthood. Gamma EEG activity begins to appear at about 16 months in humans and continues to be more and more prominent until early adulthood. On the other hand, EEG slow waves in the delta and theta bands become less prominent in the resting and task-related EEG across the same developmental period in humans. Longer range EEG synchrony as measured by coherence becomes more pronounced as the human brain develops starting in childhood through early adulthood. Because these EEG synchrony changes are mirrored by drops in synaptic density from infancy to adulthood, this tendency to see more gamma and more long-range synchrony has been suggested to reflect the mechanism of plasticity-dependent pruning.[8]

In humans, the initial trajectory of these natural selection neuroplasticity mechanisms is toward much greater intelligence through infancy, child development, and adolescence. Human intelligence as it is currently measured then levels off toward early adulthood and declines very slowly thereafter. While cerebral cortical thickness in development may be influenced by processes such as the proliferation of myelin which is the substance that typically surrounds axons, it also may correlate to usage-dependent synaptic plasticity related to sprouting and pruning. The cross-sectional results show an initial relatively negative correlation in very early childhood between cerebral thickness and intelligence in most areas. But this relationship reverses in later childhood and becomes sharply positive in many of the same regions. Given that developmental cortical changes in childhood largely seem to reflect plasticity, the interpretation is that a more plastic developing brain that shows the greatest relative degree of synaptic pruning and sprouting in the trajectory between early and late childhood will be a more intelligent brain.[9] Eventually in adolescence and early adulthood, the cross-sectional correlation between neocortex thickness and intelligence flattens a bit, but remains positive. More intelligent adults do tend to have a thicker cortex.[10]

While reasoning and verbal intelligence scores are known to be correlated, these two aspects of intelligence reflect separable fluid and crystallized intelligence systems that are presumed to have a separate neural basis. This is because fluid intelligence or reasoning seems to depend upon working memory, whereas crystallized intelligence is closely connected to accumulated semantic knowledge in long-term storage in what may be termed verbal intelligence. This separation between fluid and crystallized intelligence is also supported by evidence that cortical thickness in left hemispheric neocortex structures centered in the lateral temporal lobe is correlated to crystallized intelligence in young adults. In contrast, bilateral increases in brain activation are observed during a reasoning task in frontoparietal structures in correlation to higher fluid intelligence.[11] The same frontoparietal structures are known to be important in working memory functions that include executive control. This fluid aspect of intelligence that controls our deep reasoning ability shows more rapid decline than our crystallized or automatic intelligence abilities through normal aging. In parallel to these cognitive changes, the gray matter density of frontal and parietal regions shows more marked decline between childhood and early old age than the gray matter density of lateral temporal lobe neocortex regions, although myelin changes may be at work to some extent in these changes earlier in life.[12] Yet, crystallized verbal abilities such as verbal fluency clearly show average level declines in normal elderly in both cross-sectional and longitudinal studies that attempt to separate preclinical Alzheimer's cases from noncases. So, there is relatively greater average loss in fluid abilities, but crystallized abilities also show significant average age-related loss. Even if a person is not on a trajectory to show cognitive dysfunction consistent with dementia at a very old age, they are at least likely to show some loss of abilities with older age. An important point though is that these studies simply show average trends, but there are vast individual differences, so certain people are much less likely to show significant decline with aging.

Although the effect of aging is different in different brain structures and even within different parts of the same structure, hippocampus and neocortical circuits seem especially vulnerable to neuroplasticity deficits. Still, with notable exceptions so far restricted to an isolated prefrontal cortex area involved in working memory,[13] substantial neural loss related to aging has not been observed in controlled studies that excluded diseased cases. On the other hand in Alzheimer's disease cases, widespread neural loss is observed in full-blown dementia.[14] While widespread neural loss is not observed in normal nondiseased brain aging, other noticeable changes are observed in normal aging. Significant age-related deficits are seen in long-term potentiation (LTP) and long-term depression (LDP) in brain areas that include the hippocampus.[15] There is also evidence that decreases in dendritic spine branching size and numbers are associated with aging in cerebral cortex in primates.[16] Thus, the most noticeable age-associated impairments in nondemented older humans may be relatively restricted to synaptic signaling mechanisms in neocortical and hippocampus circuits, with the prefrontal cortex especially showing signs of deficits. This would be expected to lessen the ability to learn new explicit memories through the natural selection neuroplasticity mechanism in working memory, but the practical extent of such impairment may depend upon many factors other than chronological age.

2. BRAIN AND COGNITIVE CHANGES IN VERY EARLY ALZHEIMER'S DISEASE

A continuing and important question concerns whether Alzheimer's disease may be the end result of normal age-related cognitive decline or whether it is truly a separate disease process. The reason that this is such an important question is that the prevalence of dementia consistent with Alzheimer's disease is so high with advanced age. In fact, one very influential study known as the East Boston study put that prevalence estimate at 50% at age 85 and older with at least very mild dementia consistent with probable Alzheimer's disease. Yet, even though the prevalence of clinically diagnosed probable Alzheimer's type dementia is very high in these studies such as East Boston, there is still a substantial percentage of the oldest old who do not show evidence of cognitive impairment consistent with preclinical Alzheimer's disease or dementia. Therefore, Alzheimer's disease does not seem to be a necessary outcome of aging, but it is an outcome that has a substantial nontrivial chance of happening with very advanced age. Because it can and is avoided so often though, it must be assumed that it has causes that are distinct from the cognitive decline that is associated with normal aging. While advanced age does appear to be a major putative causal risk factor, it is unlikely that age is even the most important causal factor in Alzheimer's disease. What then is distinct about the cognitive and neural deficit in very early preclinical Alzheimer's disease compared to normal aging and what may cause this deficit?

Studies now consistently suggest that the earliest and strongest deficit in what is now called preclinical Alzheimer's disease may be within the medial temporal lobe structures including entorhinal cortex and involving hippo-campus.[17,18] Temporal lobe abnormalities in mildly demented probable Alzheimer's began to be seen in topographical scalp EEG imaging studies in the mid-to-late 1980s and early 1990s.[19,20] What was striking about the finding of EEG temporal lobe slowing in Alzheimer's type dementia from that era was that a similar though less severe pattern of focal temporal lobe EEG slow activity had been reported in roughly about 25% of nondemented elderly at age 70 since the 1950s. There had been many attempts over the years prior to the early 1990s to identify a cognitive deficit associated with this EEG temporal lobe slowing in nondemented elderly, almost all efforts had been unsuccessful and none had identified a recent memory deficit like would be expected in very early preclinical Alzheimer's disease. The only effect that had been reported was a minor verbal fluency deficit, but this was not the forgetfulness memory deficit that would be expected in Alzheimer's disease.[21] As reviewed above, verbal fluency deficits are characteristic of normal aging and are not the hallmark recent memory problem that most would believe to be seen in Alzheimer's disease.

In 1991, we reported that nondemented elderly with focal left temporal EEG slowing had deficits in a verbal recent memory test.[22] This test is a part of the Wechsler Logical Memory test typically used in assessing recent memory performance in Alzheimer's disease. An example of this test is provided in the next section. We found that those with left temporal slowing forgot more items over the course of a half hour period. That is, when the delayed recall scores were subtracted from the immediate recall scores to give a measure of forgetting, a fairly large forgetting deficit was found in the proportion of our sample with the left temporal lobe slowing. In all other ways across a fairly wide spectrum of mental tests that included immediate explicit memory, intelligence tests, implicit memory, verbal fluency and even personality tests, these individuals with left temporal EEG were, on average, no different from those without left temporal slowing. We interpreted our results to indicate that left temporal lobe EEG slowing was most likely an early preclinical sign of Alzheimer's disease. Because the prevalence of clinically diagnosed probable Alzheimer's disease at age 80 was estimated to be roughly 25% in the East Boston Study's estimates that included community elderly, we estimated that the left temporal lobe EEG slowing was present at least 10 years prior to the clinical diagnosis of probable Alzheimer's disease.

While the EEG as a measure of brain abnormalities had the advantage of a long history of data that characterized the properties of this temporal lobe slowing, it has the disadvantage that it has very poor spatial resolution and one can only guess about the neural source. Because of the far field volume conductive effects of electromagnetic signals, a strong source in the medial part of the brain may still exhibit a strong signal at the scalp. The best estimate for the source of this EEG scalp recorded temporal lobe slowing related to the recent memory deficit in the elderly was that it should have a medial temporal lobe origin either directly by reflecting that brain area or indirectly by reflecting brain areas that closely connected to the medial temporal lobe. This interpretation is based upon neuropsychological work by Brenda Milner et al. who had shown that patients such as H.M. with bilateral medial temporal lobe lesions including the hippocampus had specific deficits where they had normal immediate memory, but extremely poor or nonexistent delayed recall. Just like with our group of nondemented elderly with EEG temporal lobe slowing, H.M. had no other obvious cognitive or intellectual deficits.

The story that emerged from our work is that Alzheimer's disease begins with a very focal abnormality in the temporal lobe memory system involving the entorhinal cortex and hippocampus and then gradually spreads to other areas of the brain when widespread cognitive decline that is dementia is expressed. This process may take an average of 10 years or longer to unfold from the beginning stages of a focal recent memory problem to the clinical diagnosis of dementia consistent with Alzheimer's disease. This basic story now seems to have been supported widely through the use of other brain imaging modalities. For example, the recent memory deficit in preclinical and clinical Alzheimer's disease is associated with atrophy in the medial temporal lobe that includes the hippocampus.[23] For example, longitudinal study of the MRI[24] shows that hippocampal measures, especially hippocampal atrophy rate, best characterize mild age-associated memory impairment that likely would be called preclinical Alzheimer's disease. Yet, global brain atrophy discriminates Alzheimer dementia from this preclinical, predementia Alzheimer's group. The most striking finding here is that regional measures of hippocampal atrophy are the strongest predictors of progression from mild cognitive impairment to dementia consistent with Alzheimer's disease.[25] In another study, medial temporal lobe atrophy distinguished probable Alzheimer disease and amnesic mild cognitive impairment subjects that are likely preclinical Alzheimer's from older adults with nonamnesic mild cognitive impairment or no cognitive impairment subjects. In the same study, medial temporal lobe atrophy helped to predict diagnosis at baseline, and predicted transition from no cognitive impairment to mild cognitive impairment and from mild cognitive impairment to probable Alzheimer's disease.[26] A similar story is told in fluoro-2-deoxy-D-glucose positron emission tomography (FDG-PET) research. That is, there is a specific medial temporal lobe hypometabolism during preclinical mild cognitive impairment while much more global hypometabolism is seen in dementia consistent with probable Alzheimer's disease as demonstrated in a comprehensive review of this literature.[27] The above cited studies that have attempted to determine the rate of conversion from preclinical non-dementia with a focal recent memory deficit to clinically diagnosed Alzheimer's have put this conversion rate at about 10% per year. In other words, this figure is roughly consistent with at least a 10 year average preclinical period which was our original estimate.

A promising tool seems to be PET imaging of the putative amyloid causal protein for diagnosis.[28] Yet, the possibility of PET imaging of amyloid as a diagnostic tool is still inconclusive, but one study suggests that there may be patterns of effects that are somewhat different from MRI hippocampal atrophy in terms of rate of decline in a 2 year follow-up study in patients with mild cognitive impairment consistent with preclinical Alzheimer's disease.[29] Whether the PET imaging of amyloid will be as useful as MRI and FDG-PET studies in the study and diagnosis of preclinical Alzheimer's disease remains an open question. The possibility of something more than just a clinical diagnosis of preclinical or probable Alzheimer's disease that is closer to the confirmatory neuropathological diagnosis seems to be a good possibility perhaps in the near future, but greater understanding is needed for the accuracy of the amyloid PET imaging technology. In any case, in 2011 the National Institute on Aging and Alzheimer's Association Work Group on the Clinical Diagnosis of Alzheimer's Disease issued a completely new recommendation. This recommendation was that preclinical Alzheimer's disease should be recognized now for research purposes as a separate diagnostic category because this category will be useful for example in pharmaceutical research studies. Brain imaging measures such as PET and MRI were mentioned in this workgroup as being very important tools in this diagnosis. One major benefit of this early preclinical diagnosis category may be better pharmaceutical trials that could point to preventive interventions for memory loss and cognitive decline consistent with Alzheimer's disease. Indeed, a recent clinical trial suggests that positive long lasting memory enhancement benefits are observed in response to vitamin B12 in the age-associated memory impairment consistent with preclinical Alzheimer's, so getting to the root cause of the medial temporal lobe episodic memory disturbance in Alzheimer's disease may pay huge dividends. Other promising pharmaceutical trial results also have been reported.[30]

3. A RELR MODEL OF RECENT EPISODIC AND SEMANTIC MEMORY

Some better understanding of the connection between episodic and se-mantic memory, along with the disturbances seen in preclinical Alzheimer's disease, can be seen through a memory test similar to the Wechsler Logical Memory[31] test used in very early Alzheimer's disease. A patient is initially read a news story paragraph and told to try to remember as much as possible because they will be asked what they remember immediately after the story is read. Once the story is read and the patient states what they remember, they are then told to remember as much as possible because they will be tested again later. The delayed recall test is then 30 min later, but the patient is completely occupied with other cognitive and memory tests during that 30 min period so they cannot simply rehearse to gain good delayed recall memory performance. An example of the type of news story that could be used is the following story:

> *The American Liner New York struck a mine near Liverpool Monday evening. In spite of a blinding snow storm and darkness, the sixty passengers including 18 women were all rescued though the boats were tossed about like corks in the heavy sea. They were brought into port the next day by a British steamer.*[32]

Immediate recall may be achieved by focusing on or even rehearsing the meaningful semantic associations within news stories like these when they appear in working memory. These semantic associations are previously learned predictive models that essentially serve to chunk the many potential phonemic feature combinations into smaller sets that uniquely predict meaningful concepts. These meaningful concepts then become the pre-dictive and explanatory features in the new learning episodes. This strategy is depicted in Fig. 7.1 which breaks the memory for the New York Cruise

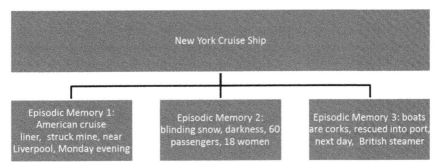

Figure 7.1 New York Cruise Ship Episodic and semantic memory.

Ship into three distinct episodes which would be processed sequentially in working memory in the same order in which they appear in this story. The features in these episodes are previously learned semantic memories like "Liverpool", "blinding snow", "steamer", etc. Each of the three episodic memories in working memory is then a unique association of these previously learned semantic concepts.

Notice that the larger memory here about the American cruise ship New York is really both a semantic memory and an episodic memory. That is, it is the basis of the concept and it is a sequence of events in time that would be described as a set of episodes. Many semantic memories also contain episodic events as predictive and explanatory elements. For example, most people have a semantic memory of John F. Kennedy that probably contains an episodic memory involving the assassination as one explanatory feature, as there is a tendency to remember episodes that have substantial emotional associations. Still, as people are exposed to a semantic concept across many different episodes, they tend not to remember episodes anymore as defining features. Even when people do retain episodes as defining features, they may no longer retain the temporal order. Semantic memories then can be viewed as predictive models that are often built from episodic information as the basic features; but much of the detail such as temporal order of events may be lost at the level of the semantic memory.

Table 7.1 shows what training data for a RELR model that codes these three episodes look like. The features are the semantic features which are assumed to have been previously learned, but a feature like "blind snow" also could be broken into separate features. There is no feature here to represent the temporal order of the episodes, but that could be easily included. Each episode could be imagined to be processed sequentially in a working memory process. All observations within an episode are assumed to be independent in the sense that they do not cause or depend upon other observations in that episode as may happen in an idealized modal or most likely neuron in a neural ensemble. Table 7.2 shows the posterior memory weights that result from such RELR models when there are only 10 training observations as in Table 7.1, and it also shows how these weights increase with more training observations. With 760 observations, there is significantly greater weight than with 190 observations which have greater weight than 10 observations. If one assumes that neurons that code working memory roughly average 100 spikes/s and that these working memory events roughly average 2–4 s, then these data give some idea of how increased exposure to these episodes through increased time or rehearsal in

Table 7.1 Training of memory for New York Cruise Ship. Each of the three episodes has 10 training observations shown as rows with the features shown as columns. Those cells that are binary coded as "1" reflect information in the news story about the cruise ship New York. Those that are binary coded as "0" could reflect other features and episodes that are related to New York and cruise ship, such as Manhattan, 9/11, Twin Towers, Wall Street, Carnival, Titanic, etc. that may be spontaneously recalled in connection to hearing a story about a cruise ship named New York.

Ship	Struck mine	Near liverpool	Monday evening	Episode 1 output
1	1	1	1	1
0	0	0	0	0
1	1	1	1	1
0	0	0	0	0
1	1	1	1	1
0	0	0	0	0
1	1	1	1	1
0	0	0	0	0
1	1	1	1	1
0	0	0	0	0

Blinding snow	Darknes	18 women	60 total	Episode 2 output
1	1	1	1	1
0	0	0	0	0
1	1	1	1	1
0	0	0	0	0
1	1	1	1	1
0	0	0	0	0
1	1	1	1	1
0	0	0	0	0
1	1	1	1	1
0	0	0	0	0

Boats are corks	Rescued into port	next day	British steamer	Episode 3 output
1	1	1	1	1
0	0	0	0	0
1	1	1	1	1
0	0	0	0	0
1	1	1	1	1
0	0	0	0	0
1	1	1	1	1
0	0	0	0	0
1	1	1	1	1
0	0	0	0	0

Table 7.2 RELR's posterior memory weights for features related to New York Cruise Ship with balanced target/nontarget sample for learning episodes as shown in Table 7.1 with 10, 190, and 760 training observations per episode.[33]

All features-10 training observations per episode
3.597

All features-190 training observations per episode
4.116

All features-760 training observations per episode
4.557

working memory may translate into greater potentiation and RELR regression weights. All posterior weights for all features within a given training observations per episode condition are the same. This is because no prior weights were imposed in this simulation and because all features were exposed to the same amount of training within each episodic condition. Obviously, differential posterior weights would result if differential prior weights were imposed and if differential memory practice had been allowed across features.

Although this New York cruise ship simulation did not introduce diversity into the time allocated to the three different episodes and then select the most important features through Explicit RELR, it is obvious that those features that received the most exposure would most likely to be selected as parsimonious features. Similarly, when differential exposure time is allowed for different episodes as should occur in the brain, features in the first episode should receive more exposure and be more likely to be recalled in a parsimonious representation. This is because there is more time for attention or rehearsal compared to the second and third episodes. In fact, substantial empirical research suggests such a primacy effect in tests of immediate recall from working memory, as, for example, earlier items in a list of words are recalled more readily. The same line of research also suggests a recency effect, as later items are also more readily recalled. The recency effect is explained as being due to later items still being in working memory at the time that immediate recall is tested.[34]

Delayed recall is obviously a much more difficult task than immediate recall. It also seems that the optimal strategy for good delayed recall is likely different from that used in immediate recall. Whereas reliance upon shallow repetitive rehearsal could be somewhat effective in immediate recall, effective delayed recall likely requires more reliance on deeper processing such as elaborating on the meaning of each episode. As examples of the deeper associations in the New York cruise ship story, the first episode might be remindful of an episodic memory for another ship striking a mine from WWII. Semantic relationships between features and episodes also could be elaborated upon, like Liverpool giving an image of a "pool" or water or like corks bobbing up and down in this pool of water. With such deeper elaborative processing that serves to chunk smaller episodic features into meaningful concepts and images, there will be a greater likelihood of successful delayed recall.

Whatever strategy is used for deeper processing, the delayed recall is expected to be more parsimonious but less accurate than the immediate recall results and very susceptible to missing features which can cause a failure to trigger an entire episode. Similarly, Explicit RELR also tends to return representations that are parsimonious and very susceptible to recall failure due to missing features at the time of recall in a scoring model. Older adults who are likely to be at the earliest preclinical stage of Alzheimer's disease show a much less substantial primacy effect in recalling words from a list.[35] Explicit RELR also will show a diminished primacy effect if there are many missing features or missing values due to degraded information as expected at delayed recall, which is very much unlike Implicit RELR's diffuse processing which is much less affected by missing information at recall.

The brain's explicit memory processing seems to reconstruct an episode of previous experience or the meaning of a scene or word in terms of a small number of associated explanatory elements. This seems to involve very specific and parsimonious reciprocal feedback signals connecting medial temporal lobe structures including the hippocampus to neocortical association structures. The dependence upon the hippocampus or associated medial temporal lobe structures is especially important when that memory is from the more recent past. Like Explicit RELR, this type of explanatory model is good at explanations of what has been learned in the past, as when people remember the parsimonious set of features in a news story. Yet this type of "explanatory model" is very susceptible to the forgetting that is observed early in Alzheimer's disease, whereas implicit memories are not as affected by early Alzheimer's disease. These Alzheimer's explicit memory deficits may be due to the effect of information loss in such parsimonious memory representations. So the loss of input features in the brain's recent explicit memory system may have a more exaggerated effect on performance than any similar loss in more redundant implicit memory representations. This is analogous to how the loss of input feature information in Explicit RELR through missing values or variables would have a much stronger effect on predictive performance than a similar loss in Implicit RELR.

Both animal and human evidence show that the hippocampus is involved in parsing perceptual features into separate episodic events and then storing, maintaining or recalling these episodic sequences events in time.[36,37] Different parts of hippocampus may even exhibit differentiation in terms of the time duration between events that can be optimally processed and remembered as a sequence.[38] Thus, it is expected that mechanisms that parse features in the world into separate episodic events and remember the causal sequences to be quite complex in the brain. Deep processing may involve the hippocampus in computing likely causal explanatory memory models that string together stable and parsimonious features in separate episodic events into an appropriate temporal sequence that is itself a stable and parsimonious representation. This also may involve chunking information into a parsimonious set of explicit memory features, concepts and images that are much more compact and meaningful than diffuse implicit memory representations. In any case, Explicit RELR may be helpful in understanding how a very parsimonious representation of separate episodic events consistent with explanatory and causal inference can be computed.

It is interesting that remote semantic memory does not appear to be greatly impaired in early preclinical Alzheimer's disease relative to elderly controls. If semantic memories like the New York Cruise ship memory are developed from repeated new episodic learning in analogy with this RELR model, then with enough varied episodes over long periods such as years, the remote semantic memory also could be a more diffuse memory representation composed of many different features. So, the distinction between explicit and implicit memory may be clouded in the case of the remote semantic memories where the brain no longer may have just a few features which characterize the memory; this would protect such memories from erosion earlier in the course of Alzheimer's disease.

4. WHAT CAUSES THE MEDIAL TEMPORAL LOBE DISTURBANCE IN EARLY ALZHEIMER'S?

Even though there is now a reported trend for small average deficits in fluid reasoning abilities in predementia Alzheimer's disease in larger studies, these differences are too small to be separated from age-related drops in fluid abilities.[39] Very early preclinical Alzheimer's patients are clearly not yet demented. Their reasonably normal immediate recall performance indicates working memory processing that is still as functional as age-matched controls. Digit span, which is the ability to rehearse digits phonologically in working memory similar to a phone number, was also not different in our study of temporal lobe slowing.[40] Thus, it is expected that shallow working memory processing like phonological rehearsal is not greatly impaired in very early predementia Alzheimer's disease compared to age-matched nondemented controls. Yet, the process that stores working memory information into more long-term memory is greatly impaired.

So what goes awry at the origin of Alzheimer's disease that disrupts the exchange of information between working memory and a longer term explicit memory store that exists outside of the medial temporal lobe? At a neural level, the obvious candidates are the repetition-based LTP and LDP deficiencies in the hippocampus and other related medial temporal lobe structures, as the half-hour delay would seem to be too short of a period for structural neural plasticity changes related to sprouting and pruning synapses. At a cognitive level, there is obviously a deficiency in storing or retrieving new explicit memories, but it does not seem to be related to a more shallow phonological rehearsal ability deficit as digit span is not deficient.

Deep processing involves the ability to learn and understand new associations between separate episodic events and semantic concepts including the causal connections based upon working memory associations. Compared to age-matched controls, predementia Alzheimer's patients still may be able to repeat the story when they initially hear it because shallow processing such as the ability to rehearse a phone number is not initially impaired as H.M. also can perform such a task. People with good delayed recall ability are able to do much more than just repeat the story in rote recall in at least some details. That is, they are able to retain an understanding of the deep connections in the story in an abstracted form that chunks specific details into a parsimonious set of higher level concepts or associations. The medial temporal lobe impairment may disrupt this deep processing ability that allows the understanding of new connections in recent experience.

How the EEG slowing that gives a greater predominance of delta and theta rhythms may relate to this preclinical Alzheimer's recent memory impairment is not understood. Yet, evidence was reviewed in the last chapter that theta rhythms are potentially involved in timing of computational events that transfer information from working memory to explicit long-term memory. This research is consistent with the possibility that the theta rhythm reflects the oscillating maintenance of information that is necessary to store the temporal order of episodic events through the hippocampus. Any disruption in the timing of such critical oscillating neural synchrony may lead to a disruption in the likelihood that events could be stored or retrieved later. Such a disruption in timing naturally would be associated with decrements in LTP, LDP, and neural plasticity. Consistent with this possibility is evidence from implanted electrodes in epileptic patients that theta rhythms that reflect feedback between prefrontal and medial temporal cortex become much more coherent during delayed recall of words compared to a baseline condition, which would be consistent with recall critically depending upon the synchrony of computations in these structures.[41] Also consistent with this possibility is the observation that human memory is strongest when there is a strong coupling between local theta oscillations and neuronal spiking during learning in medial temporal lobe structures including the hippocampus.[42]

Most of the studies that have been concerned about risk factors for Alzheimer's disease are correlation studies which have the limitations of data mining predictive modeling that has been reviewed through this book. Another problem is that very few of these studies have been concerned with risk factors for preclinical Alzheimer's disease, so the comparisons that involve probable Alzheimer's disease and "normal elderly controls" contain a high percentage of preclinical patients in the control group which will wash out and potentially obscure important risk factors. Thus, it is very difficult at present to make any strong claims about risk factors specifically for Alzheimer's disease. In any case, there are well-known experimentally studied factors that seem to protect against cognitive decline in aging and these seem to be the same factors that promote neural plasticity.[43] These include relatively high levels of early environmental enrichment or education, a low-fat and high-antioxidant diet, aerobic exercise, and continued novel and challenging environments during aging, as they all may be linked to enhanced brain activity and neural plasticity in medial temporal lobe structures assumed to go awry at the origin of Alzheimer's disease.[44]

Yet, the best approach to promote neural plasticity in aging may be a combination of these behavioral factors. As an example, one experiment showed that old dogs given both a high-antioxidant diet and an enriched housing with other dogs, toys and outside walks showed the best cognitive performance compared to dogs who received no beneficial diet or environmental treatment or who only received one treatment.[45] There is similar evidence in older humans that a combination of aerobic exercise and cognitive training may improve recent memory more than either treatment individually.[46] However, a very large study that just looked at cognitive training alone did not report substantial transfer of improvements to other tasks in older adults.[47] Another smaller study also recently reported that cognitive training improvements alone do not transfer to other cognitive tasks, but this study showed that a regimen of viewing documentaries about the real-world did produce improvements that transfer to other cognitive tasks.[48] So while the benefits of artificial cognitive training are questionable, a real-world activity like viewing an intelligent documentary that promotes deep processing might help prevent memory loss and cognitive decline consistent with Alzheimer's disease.

If a mechanism like Explicit RELR is able to lay down the posterior memory weights for explicit memories, then the parsimonious features, episodes and concepts that are learned would be especially susceptible to the neuronal and synaptic loss that occurs in very early Alzheimer's disease. People with more education are less prone to age-related cognitive decline and Alzheimer's disease, but the mechanism predicted by Explicit RELR would be that they have more extensive and richer semantic networks that also serve to create deeper and more abundant initial candidate inputs for working memory and fluid reasoning abilities. Biological factors such as the known AP04 genetic risk factor and previous head injury also may impact on this mechanism either by directly causing extensive or rapid neuronal loss, or by other indirect mechanisms such as metabolic deficiencies. Diet and exercise promote better brain metabolism and this would serve to help the energy exhausting Explicit RELR working memory process and also prevent age-related synaptic efficiency losses. Plasticity is still present in old age with novel learning situations, so this would predict that older adults who continue to engage in new learning as in watching intelligent documentaries frequently also may be able to help prevent cognitive aging and Alzheimer's disease. This is entirely consistent with Darwin's natural selection concept that cognitive abilities will atrophy if they are not exercised.

5. THE MIND–BRAIN PROBLEM

The complete breakdown of a mind in more severe Alzheimer's disease is a very troubling experience for anyone who has experienced such an event in a close relative or loved one. This experience calls into the question the nature of human existence and how the essence of our mind, our conscious cognition, arises from the brain. This mind–brain problem, or mind–body problem as it is known in philosophy, is the most significant problem in cognitive neuroscience including applied areas like machine learning. Descartes' famous suggestion that *Cogito Ergo Sum* or "I think therefore I am" presupposes that a thinking mechanism is the basis of our existence. Any general neural theory of cognition must come to grips with the full implications of this problem. Machine learning also must deal with this problem, as there may be significant limitations in how artificial intelligence could ever approach the brain's capability to provide a unified conscious experience.

Descartes claimed that the brain's pineal gland is "the principal seat of the soul, and the place in which all our thoughts are formed".[49] Descartes localized what we may call mind in the pineal gland because it is a unitary structure that exists in-between the two hemispheres of the brain and so is often referred to metaphorically as the "third eye". While Descartes may have been wrong in the specifics of where mind could be located, his idea that mind implies consciousness and so it must be based upon that which unifies our experience still remains very central in all influential modern ideas about the relationship between brain and consciousness. What then could serve to harmonize our brain's processing into a unitary whole that is the essence of consciousness?

The answer that is proposed in this book is related to the finding that field potentials can change dynamic firing properties of spiking neurons.[50] Such an interaction between field potentials and spiking neurons would be expected in both unconscious, implicit and conscious, explicit brain processing. Yet, the present view is that consciousness is reflected by synchronous and oscillating neural ensemble field potentials that form stable standing waves that are the essence of stable information. So highly organized field potentials that form standing waves reflect consciousness and can change dynamic firing properties of spiking neurons just as spiking neurons can change field potentials. If ghost is taken here to mean a mechanism that is not understood, the present notion specifies the physical manifestation of that ghost at least in a way that explains how spiking outputs in neurons may not completely depend upon input features. This stable information theory is influenced by the oscillating neural synchrony literature reviewed in the last chapter, and it is also related to but not identical to a number of other proposals for a neural basis for consciousness that is localized in brain processes.

One such related proposal is Edelman's concept of a neural basis for consciousness based upon his Darwinist theory. This theory views consciousness to arise from integrated neuronal group activity that arises from extensive reentrant feedback signals across major areas of the brain with the most important exchanges between thalamus and cerebral cortex to process incoming sensory information, along with the hippocampus to allow explicit learning.[51] However, Edelman stops short of suggesting that consciousness could influence such neural activity. In Edelman's view, consciousness is a process that seems to parallel neuronal activation. But unlike the present view which is that causality can move in either direction, Edelman suggests that consciousness cannot be a driver of neural activity. Instead, he strongly argues that all causality must come from neural activation itself, although Edelman's theory could be consistent with field potentials influencing neural activity since field potentials may be viewed as an aspect of neural activity.[52] Edelman also has opposed any suggestion that his neuronal group selection theory is consistent with the idea that the brain is a computer; the implication is that there are strong limits in the extent to which the brain also could be mimicked by a computer. Yet, Edelman obviously believes that neurons are capable of computation given that he has published extensively on computation models of neuronal groups.

Edelman's ideas have had a large influence in neuroscience including direct influence in terms of his collaboration with Giulio Tonini who is known for the integrated information theory of consciousness. This theory is that consciousness is a graded entity whereby higher levels of consciousness correspond to greater integrated information and less relative entropy. Tonini's view would not even necessarily localize consciousness in brains and instead would suggest that consciousness could exist in any system which has the capability of allowing high information patterns of activity to interact. Thus, in Tonini's view, consciousness could exist in silicon chips and computers. However, Tonini also suggests that the corticothalamic feedback signals in the mammalian brain are based upon the precise architecture necessary for the integration of information that is the basis of consciousness. Oscillating neural synchrony is associated with consciousness in this view because it has low relative entropy and is thus associated with a large amount of integrated information.

E. Roy John's theory has been heavily influenced by Edelman and Tonini. This theory is that consciousness is a specific emergent property of the brain's local oscillatory electromagnetic fields. John argues that the totality of such nonrandom coherent oscillatory local field signaling in the brain is consciousness. As with Tonini, John's view is that nonrandomness or low relative entropy or information that he calls negative entropy is the defining characteristic of consciousness in these signals. Yet, John suggests that such consciousness is necessarily a genetic property of neural systems, so the implication is that it would not be possible in computers. McFadden[53] also has proposed a neuroelectromagnetic field theory of consciousness. But he goes one step beyond John and suggests that the synchronized local field potentials actually can influence the probability of binary spiking in neurons. Thus, oscillating neural synchrony is not just viewed to be an epiphenomenon of cognitive function in McFadden's view, but instead oscillating neural synchrony would be able to cause changes in cognitive function. This ability for oscillating neural synchrony to influence axonal spiking is assumed to be the basis of the subjective sense of free will although this is a determined free will based upon a physical mechanism.

The McFadden theory is very similar to the present stable information theory. The exception though is that the present theory is agnostic on whether free will is deterministic or not, along with the exact mechanism of how consciousness might arise as an emergent property of oscillating neural synchrony. This ambiguity should not be troubling because physics has yet to arrive at a generally agreed upon mechanism for gravity many hundreds of years after Newton said that he does not hypothesize about the mechanism of gravity. But, just like gravity or whatever mechanism gives rise to gravity is a necessary property of nature in order to get reasonable causal explanations in the physical macro world, consciousness and its ability to exert influence on neural computations might be considered to be a necessary property of the brain. Still, there are some interesting speculative mechanisms for free will and consciousness. For example, a recent book by Tse argues that the mechanism for conscious free will is in the rapid altering of synaptic connection weights in the brain. In this way, conscious free will cannot change the present but it can change the course of actions in the future. In essence, this mechanism is simply that free will is able to play a hand in the rewiring of the brain.[54]

Another proposed mechanism concerns how consciousness may be an emergent property of oscillating neural synchrony in the gamma frequency range. Hameroff and Penrose suggest that tiny structures that are known as *microtubules* in dendrites are the basis of quantum interactions that give rise to consciousness. In quantum physics, quantum entanglement effects are observed when two particles interact physically to exhibit synchronous behavior in momentum, spin, and polarization and also exhibit such synchronous behavior when they are separated. This quantum model of consciousness has resulted in significant academic research and dialogue about whether or not it is feasible from a quantum physics perspective.[55] Hameroff proposes that this action-at-a-distance entanglement of spatially separated quantum particles would be partially involved in the mechanism that produces synchronous action-at-a-distance EEG gamma oscillatory effects in the brain.

Yet, Hameroff also suggests that known classical ways for signals to move across neurons through tight gap junctions also would play a central role in how gamma oscillations move so rapidly throughout the brain to create long-distance synchrony effects. Hameroff points out that this possibility that the gamma oscillations that are the basis of consciousness can move globally across brain tissue through nonsynaptic gap junctions is a return to Golgi's nineteenth century idea of the Reticulum. Thus, this oscillating gamma synchrony as the manifestation of consciousness represents a mechanism that moves around the brain and takes control of nonconscious activities which are controlled by classical discrete neuronal circuitry involving synapse.[56] Hameroff's work is both very creative and intriguing, but more empirical research is necessary to determine whether a quantum mechanism is the basis of high-frequency gamma electric field fluctuations and consciousness.

The mechanism that allows consciousness to arise from oscillating neural synchrony probably will not be experimentally tested and fully understood any time soon. Yet, even without a mechanistic understanding of consciousness, the goal of artificial intelligence can be viewed as allowing an extension to the brain that will make humans much smarter.[57] This extension need not be a physical interface, but instead simply could be a massive external computer. Under this present view, there is no need to think about developing computers that exhibit consciousness, as human consciousness and judgment will be needed to interpret the explanations that cognitive machines might offer. Thus, the fact that the mechanism of consciousness is not understood should not be a hindrance. This is because the creation of conscious cognitive machines is probably not a realistic goal. Instead, a realistic goal might be to use neural computation mechanisms to produce cognitive machines that aid in causal discovery and prediction.

CHAPTER 8

Let Us Calculate

"The only way to rectify our reasonings is to make them as tangible as those of the Mathematicians, so that we can find our error at a glance, and when there are disputes among persons, we can simply say: Let us calculate, without further ado, to see who is right."

Gottfried Leibniz, The Art of Discovery, 1685.[1]

Contents

This book started with the Leibniz goal of a *Calculus Ratiocinator* as a smart machine that avoids problems related to cognitive and data sampling bias and error as inputs in human decisions. At the outset, it was suggested that the *Calculus Ratiocinator* would need to model the brain's attention, memory, and reasoning processes. The RELR methodology and its ability to compute accurate probabilities based upon high-dimensional and small sample data has been proposed to be the basis of such a *Calculus Ratiocinator* model. Through the book, it has been argued that the same basic RELR model of computation at the neural level can be interpreted as reflecting cognitive computation at the neural ensemble level. Basic RELR neural computation mechanisms involving implicit highly parallel feed-forward and explicit sequential feedback processes thus followed. A very simple model of sequential online learning also followed, along with a model of matched sample causal learning based upon the interplay of these implicit and explicit RELR neural computation methods. This also led to a sense that oscillating neural ensembles that are synchronized in orchestral computations are the basis of the conscious intelligence that breaks down in Alzheimer's disease.

The central core of this theory of neural computation is a connection to information theory through logistic regression. The concern is with nonarbitrary, optimal predictive and explanatory models that are stable and more likely to replicate across independent data and modelers because they account for errors-in-variables. This logistic regression-based theory of neural computation only includes stable and non-arbitrary aspects of information theory that are also likely to be neuromorphic. It is noteworthy that various aspects of information theory more generally are not a component of the present Calculus of Thought. This includes concepts that are not easily computable or that do yet not have well agreed upon definitions or multivariate forms. For this reason, concepts like Kolmogorov complexity, minimum description length, and mutual information do not yet have a place in the present information theory of neural computation, although that could change in the future with more agreement in how to compute these concepts and if they can be shown to offer unique insights into neural computation.[2] The present information theory of neural computation provides a unifying view of probabilities computed in logistic regression that synthesizes frequentist and objective Bayesian notions. In terms of neural computation, prior probabilities are simply memory weights that must be influenced by previous learning including instinctual learning that has origins in the history of evolution.

RELR's two basic implicit and explicit processing mechanisms model what may be described as more automatic, shallow, fast versus more effortful, deep, slow neural ensemble computations. As a neural ensemble model of computation, RELR is a skeletal outline but it does overcome problems with previous neural computation methods that do not allow small sample learning and have problems with high-dimensional multicollinear inputs. Additionally, RELR exhibits learning-related enhancement effects of memory weights that are also found in neurons and neural ensembles.[3] Because it is a theory of neural computation that is consistent with known computational properties of oscillating neural synchrony, RELR's basis in information theory provides a common framework that will allow neuroscientists and machine learning scientists to speak in the same language. This is especially important in an era of Big Data where extremely high-dimensional computational processing is possible, so models of how neural ensembles handle such enormous Big Data computations in realistic rapid computations may be produced. Any and all mechanisms in RELR's information theory of neural computation may be eventually falsified with better empirical data like from the BAM project, but at least this theory presents a consistent story that can be tested empirically in coming years.

Social scientists, psychologists, economists, natural scientists, statisticians, mathematicians, management scientists, engineers, cognitive scientists, data scientists, and computer scientists who use predictive analytics in basic research or applications will not need to become neuroscientists to use and understand RELR. But they at least will have a high-dimensional Big Data analytics tool that has a basis in neural computation theory. In this respect, they also may benefit as future neuroscience learns more about the brain's computations, and the skeletal sketch provided in this book is updated and adjusted in the powerful self-correcting mechanism of science. Besides providing accurate probabilities and stable variable selections and regression weights that are much more likely to replicate independently, the RELR methodology overcomes problems with standard methods related to sequential time varying data and causal inference in observation data. These new sequential online learning and causal learning applications may be especially useful in today's big data world composed of rapid updates with massive high dimension data where there is enormous potential always for spurious predictions and explanations.

RELR is designed as a tool for statisticians, data scientists, machine learning experts, and researchers who understand all the nuances of sampling, research and analytics design, and who are subject matter experts or who have access to subject matter experts. In this regard, RELR can lead to cognitive machines that can help make smarter decisions if applied properly. RELR will not replace statisticians, data scientists, and researchers, but instead will help these professionals produce more valid and reliable predictive and explanatory models as inputs in decision making processes. These quantitative and research professionals have the background to understand that RELR predictions and explanations will be always still subject to error and bias due to poorly designed input features or research methodology or sampling. They also have the skill sets required to do everything possible to remove confounds and other data collection design problems that result from cognitive and sampling biases. Most importantly, these analytics professionals understand that the major reason that analytics fails is due to incorrect assumptions, so they always will need to be intimately involved to ensure that assumptions are correct. RELR is not designed to be used by those who do not have fundamental training in research and statistics and who wish to have a tool that democratizes analytics without involvement from research and quantitative professionals. There are simply too many things that can go wrong in predictive analytics. Even in the most automated application, it is very important that qualified professionals are closely engaged in the process. Otherwise, the decision makers who are the end users will be likely to have significant problems.

What needs to be remembered always is that decision makers only will be trusting of predictive analytics when predictions and explanations are reliable and valid. That is, arbitrary and random variation in predictions and explanations across modelers and/or data samples can create enormous problems. And the decision makers who are ultimately the customers of predictive analytics all instinctively know this. So this last chapter is an overview of real-world examples of how data sampling and human cognitive error and bias problems have led to problematic predictive models that have very much hurt the decision-making process. The focus is on how RELR cognitive machines can help to avoid these problems that that decision-making end users find unacceptable.

1. HUMAN DECISION BIAS AND THE *CALCULUS RATIOCINATOR*

The formal recognition of the preclinical Alzheimer's disease diagnostic category in 2011 by the National Institute on Aging and Alzheimer's Association Work Group on the Clinical Diagnosis of Alzheimer's Disease culminated more than 20 years of research that initially was not viewed as mainstream, as our original proposal of at least a 10-year preclinical causal degenerative period in Alzheimer's disease took many experts completely by surprise. For example, I was at an American Geriatrics Society conference in 1990 where the keynote speaker presented findings that showed that a cerebral spinal fluid marker was 100% accurate at classifying patients with clinically diagnosed dementia consistent with probable Alzheimer's disease versus nondemented elderly controls.[4] During his talk in front of a very large audience, he also mentioned that this protein test seemed to be sensitive to early Alzheimer's disease. During the question-and-answer period, I asked him a difficult question. If it is sensitive to early Alzheimer's disease, then why is the abnormality only found in demented patients? That is, one would expect that at least some nondemented elderly controls have preclinical, predementia Alzheimer's disease, so they should also test positive. He gave a very honest and candid response that he did not have an answer other than they were still learning about the properties of this biomarker test. As it turned out, even though his and many other tests over the years were very accurate at classifying Alzheimer's dementia vs. non-demented matched controls. These predictions were way too late to be useful.

Yet, this scientist's data concurred with the common belief that Alzheimer's disease did not have an onset usually until much later in life, as it was assumed that forgetfulness without corresponding dementia was a characteristic of normal aging. In addition, there was an implicit view which is also very prevalent in the data science community today that the best predictive model always has to be the most accurate model in terms of classification. This is in spite of the possibility that the accuracy in classification could be due to associations that occur as an effect of the disease rather than as a cause of it. So, his research and conclusions were really also just a reflection of this mainstream. In contrast, we did not find nearly 100% classification accuracy. But we believed that those cases that were wrong were likely the preclinical cases since they also had focal recent memory deficits. But this meant that there must be a long preclinical period. Because of this mainstream view that Alzheimer's is a late onset disease in the majority of the cases, our long preclinical hypothesis was not taken seriously. So, there was also no real recognition of the possible importance of our electroencephalography (EEG) temporal lobe slowing findings in terms of funding priorities. For example, when I proposed performing my own longitudinal study to the National Institute on Aging in the early 1990s to demonstrate that EEG temporal lobe slowing is an early 10-year preclinical indicator of Alzheimer's disease and to link it to other measures like magnetic resonance imaging, the grant proposals were not funded. At that time, I also do not think that physicians were comfortable with the idea of such a long preclinical period, even if it were true. In their defense, the problem was that even if this early preclinical hypothesis were true, there were no treatments available. After all, a diagnosis is simply a label which serves no function unless the condition can be remedied. While I understand that reasoning, I also sense that we lost about 10 years in Alzheimer's research until this issue began to be addressed again. All through this time, treatments were attempted and largely ineffective in patients who had progressed to dementia. It now seems most reasonable to believe that effective treatments and the prevention of further decline are only likely to be possible in the early preclinical, predementia period.

On his show that aired on February 6, 2012 on Bloomberg, Charlie Rose interviewed Dr Scott Small, the neurologist on the team that made the just reported recent discovery that the mouse model of Alzheimer's disease spreads from just one initial location in the medial temporal cortex—the entorhinal cortex.[5] In that interview, Dr Small articulated his vision, which is a predominant vision, that Alzheimer's disease might be similar to a neurological disease like Lou Gehrig's disease, as it starts in one location in the nervous system and spreads uncontrollably thereafter. Charlie asked Dr Small about why it took so long to realize the implications that Alzheimer's disease must be this kind of disease where ideally one must stop it prior to uncontrolled spreading. Dr Small's reply was that we were constrained by our thinking that Alzheimer's disease is part of normal aging until the 1980s and this original view biased us and slowed us down.

The story about how our cognitive biases get in the way of good decisions can be told about any complex decision process. When this occurs, the "experts" are often the last people to realize the truth and come to an appropriate decision. The great promise of a *Calculus Ratiocinator* is that it has the potential to avoid these expert biases.

2. WHEN THE EXPERTS ARE WRONG

Aspects of Daniel Kahneman's book *Thinking, Fast and Slow* have already been reviewed in relationship to implicit or fast cognition and explicit or slow cognition. As a cognitive psychologist by training, Kahneman has spent his career trying to understand how humans make decisions in practical applications such as medicine, business, and education. Much of this research focused on how humans are not always rational in how they make decisions. Much of this research also focused on debunking "experts" who claim to have prowess in predicting and explaining human behavior, such as counselors, money managers, and long-term political forecasters.

Kahneman's research has consistently shown that humans suffer from "cognitive illusions" that are similar to perceptual illusions like the Kanizsa Square.[6] One illusion that he reviews is the "halo effect" which is when people who have a desirable attribute such as physical attractiveness are ascribed high abilities in other attributes such as intelligence or athleticism. This "halo effect" illusion was a center point in the recent book and movie *Moneyball* when baseball scouts would erroneously associate player skills with good looks.

In another example of a cognitive illusion, Kahneman surveyed people about their prior beliefs about the likelihood of Nixon's success in his trip to China in the 1970s, and then surveyed the same people afterward to see the likelihood that their prior beliefs had changed. This research observed that people were very inaccurate in giving their prior beliefs when they were wrong. In other words, they have an illusion that they were right in their initial prediction.[7] Many of these cognitive illusions are related to errors in probability judgments. As an example, Kahneman did a study where he initially read his subjects a story about Linda a philosophy major who was involved in liberal activist causes. He then asked the subjects to choose the most probable statement from a list of statements about Linda. What he found was that many more people chose "Linda is a feminist bank teller" to be a more probable statement than "Linda is a bank teller".[8] Obviously, there are many more bank tellers than feminist bank tellers, so this specific judgment about probabilities is erroneous as is often the case more generally when humans make snap judgments about probabilities.

Kahneman points out that those facts that challenge basic professional assumptions and therefore threaten people's livelihood are often ignored. He describes his personal encounters with financial executives who thought that the bonuses that they gave to money managers were key drivers in performance, but were not pleased when he informed them that there is no correlation between year-over-year performances of money managers. Expert's beliefs in their own decision prowess are especially hard to change when supported by a community of like-minded individuals in a powerful professional culture. Kahneman suggests that people in such a culture believe that they are the chosen few who alone have unique access to the truth.[9]

Kahneman's work that led to the debunking of those "experts" who considered themselves capable of accurate predictions and/or explanations of human behavior was heavily influenced by the research of Paul Meehl. Meehl was a mid-twentieth century research psychologist whose work concerned the ability of counselors and therapists to predict and explain human behavioral outcomes such as school achievements. Meehl showed that these clinicians performed very poorly in these tasks. Kahneman writes that "Meehl's clinicians were not inept and their failure was not due to lack of talent. They performed poorly because they were assigned tasks that did not have a simple solution … they operated in low-validity situations that did not allow high accuracy."[10]

Kahneman's critique of expert decisions is based upon human failings in intuitive automatic judgments. Kahneman points out that you can trust experts when they have had ample time to learn regularities through practice in highly predictable environments where rapid accuracy feedback is provided. These are often the environments where automatic more implicit learning would be accurate. He suggests that firefighters, physicians, nurses are examples of those you can trust because they work in such environments. Money managers, therapists, and long-term political forecasters are examples of those that you may not trust. A telltale sign of a domain where experts should not be trusted is when the experts are inconsistent and consistently contradict themselves or each other in their predictions, explanations and associated decisions. Kahneman warns that "unreliable judgments cannot be valid predictors of anything".[11]

To avoid the inaccuracy and variability associated with expert decisions in domains where rapid automatic intuitive implicit memory processing fails, Kahneman suggests that one option is to employ simple formulas. He reviews an example where human intuition is worse than a simple predictive analytics modeling formula. This example is a fine Bourdeaux wine price forecasting model by a Princeton economist—Ashenfelter. This model employs three simple variables that relate to weather conditions that presumably have causal effects on the quality of grapes. This model provides a simple formula that is much better than wine experts at predicting future wine prices.

However, Kahneman actually urges caution in the use of automatic standard regression methods to generate simple predictive formulas. The reason is the inaccuracy of standard regression methods due to the well-known data sampling problems that are especially problematic with multi-collinear variables. Much of this book has been concerned with these same multicollinearity error problems and how RELR overcomes these problems. In comparison to regression methods, Kahneman suggests that "formulas that simply assign equal weights to all the predictors are often superior, because they are not affected by accidents of sampling".[12]

Although equal weight formulas can be superior to multiple regression, they do not solve the problem of which variables to select. Thus, modelers are still dependent upon experts to determine which variables should be selected in a simple predictive formula, or they use traditional variable selection methods like Stepwise methods, Decision Trees, LASSO, LARS, or Random Forests as is the common practice in today's predictive analytics applications. In a case like wine quality/price forecasting where causal factors related to grape quality are known, experts would be helpful to select the small set of likely causal variables. However, in many applications in business, medicine, government, and social science, often none of the important causal factors are known. In that case, both expert and traditional automatic variable selection are well known to be highly unreliable and prone to error and bias. As reviewed in the first chapter, the large MAQC II Consortium study that evaluated 30,000 predictive models provided evidence that differing variable selection across independent samples and expert modelers was the best predictor of poorly performing models.[13] This outcome is completely consistent with Kahneman's view that effective decision making cannot be based upon an unreliable process that often generates contradictory and biased decisions.

Thus, any traditional predictive modeling method that is based upon subjective and arbitrary decisions on the part of the modeler and/or that is contaminated by the accidents of data sampling should be considered to be a very ineffective and risky input to the decision-making process. For this reason alone, decision makers should consider only working with predictive modelers who can ensure that their methods are not contaminated by cognitive bias and the accidents of sampling, and this is precisely the justification for RELR. Experts still may be helpful in providing good candidate variables in these models, as long as multiple experts are used in areas where there is diversity of opinion on the important variables. That is, the goal in a more objective approach to predictive analytics like RELR is to interview as many experts as possible so as to get as much diversity of opinion as possible. In this way, the set of candidate features will contain many possible hypotheses for good causal and predictive variables, and the RELR model will be an objective selection of the most likely features given appropriate controls to ensure that sampling is not biased.

3. WHEN PREDICTIVE MODELS CRASH

One very good example of reliability and validity crashes in predictive models was made clear in a PBS Nova documentary in the spring of 2012.[14] The head of NIH Dr Francis Collins explained how he sent his genomic data to three different personal genomics firms to get his risks assessed for various diseases. The predictive profile across common diseases that he got back was mostly different across the models supplied by the three firms and even contradictory. The goal of cognitive machines that aid decision makers has to be not only to provide accurate predictions but also to provide reliable predictions that do not across change modelers. This is especially important in health decisions that individuals or doctors make on the basis of predictive analytics. Obviously, predictive analytics only will be able to lower health care costs and increase quality of care if it can become much more reliable than in this high profile case study.

There might be some debate about the extent to which failures in the automotive industry that required government bailouts in the past decade were caused by poor predictive modeling. Certainly, poor management practice had a huge impact, but it is hard to separate out predictive modeling because it had become an important input into the decision-making process. At the very least, unreliable predictions and explanations generated by predictive modeling added noise to the decision-making process, and did not provide actionable results to counteract the most important causal reason that led to the failures of General Motors and Chrysler. For example, in 2008, there was a GM television commercial ad that aired in the United States which reported that "cup holders, researchers say, is one of the top reasons women buy one car over another".[15] While it is unclear whether GM acted upon this modeling prediction and explanation, we know that they paid for this research and also did not produce high-quality small cars that should have been produced in the event that gas prices went higher.

Financial industry failures in recent years have been argued to be even more of a direct result of poor predictive modeling. In the late 1990s, the Long-Term Capital Management (LTCM) hedge fund was a quantitative trading hedge fund based upon data mining methods and regression that ran into trading problems that led to a $4.6 billion dollar bailout organized by the Federal Reserve Bank of New York.[16] Other examples of predictive modeling failures are seen in the 2007–2008 US mortgage banking crisis. Many important financial decisions were based upon a Gaussian copula method to price credit derivatives that nobody really understood. Many of these models typically had their performance tied to optimistic future forecasts of housing prices, and all ultimately failed. One exception to the widespread naïve belief in Gaussian copula as an effective predictive modeling method was Nassim Taleb, author of the very popular book *The Black Swan*.[17] Taleb was skeptical of this predictive modeling approach from the beginning and is quoted as calling it "charlatanism".[18] Like the poor performance of the credit derivatives price models in this 2008 crisis, consumer credit risk models also did not do well.[19] In 2011, another financial crisis began to occur this time centered in Europe, and this European crisis has also resulted in very expensive bailouts. One bank that originally had been bailed out in 2008 by US bailout funds to the tune of almost $60 billion US dollars actually had to be bailed out again in 2011 by the Europeans for an undisclosed sum.[20] This bank got into trouble the second time with poor decisions based upon expectations of future interest rates whether caused directly by poor predictive modeling or by decision makers who decided to produce their own forecasts because they did not trust the predictive modeling.

Obviously, there was excessive leverage and risk taking and very sloppy management in many of these financial institution disasters. In addition, there was a lack of regulation. Also, much of the risk taking, such as the easy credit for housing, was actually encouraged in the United States by the government's equal access to housing policies. Yet, like the General Motors and Chrysler inability to produce actionable predictions of demand for the types of vehicles that consumers would actually purchase, inaccurate predictive modeling was also a part of every one of the major financial institution crises or failures that occurred starting with LTCM in 1998 all the way through the continuing European crisis that began in 2011. A recent book by Emanuel Derman titled *Models Behaving Badly* argues that bad econometric modeling that assumed that human behavior was stable and predictable played a very central part in the 2008 US financial crisis that he witnessed at Goldman Sachs.[21] In fact, Derman even suggests that human behavior is ultimately impossible to model and predict because it is too varied and changes too randomly, so this pessimistic view is that these problems will continue to occur again if predictive modeling is relied upon.

4. THE PROMISE OF COGNITIVE MACHINES

Predictive models can show even greater problems than expert hypotheses because predictive models are not only subject to the accidents of sampling but are also subject to the cognitive biases of expert judgments. This possibility is suggested by the work of Ioannidis who presents a metareview that shows that data mining studies in medicine seldom replicate, whereas large randomized controlled experiments are likely to replicate. Ioannidis's results show that problems related to lack of replication are a combination of biases that investigators might have, along with measurement, sampling, and related confoundedness problems in correlation findings. While these problems can occur in randomized controlled experiments, they are just much more likely to happen in data mining work. In fact, randomized controlled experiments offer a natural way to test and guard against biased expert hypotheses. Unfortunately, predictive modeling based upon data mining actually may support these biases and lead to more disastrous decisions because there is an illusion of objectivity when expert biases are supported by a predictive model. This is because the possibility that the predictive model later could be shown to be spurious often seems to be forgotten.

Subjective bias problems might be especially problematic in academic research applications that attempt to inform public policy as occurs in the social sciences. On February 11, 2011, the New York Times carried a story about a poll conducted by University of Virginia social psychologist

Jonathan Haidt at the 2011 Society for Personality and Social Psychology Conference.[22] Dr Haidt gave a talk in front of about 1000 attendees and asked "how many considered themselves politically conservative"—about three hands went up. He also asked how many considered themselves moderate—about three dozen hands went up. In contrast, the vast majority of hands went up when he asked "how many considered themselves politically liberal". The field of social and personality psychology is involved in asking some of the most important social questions of our time related to gender differences in roles and intelligence patterns, psychosocial factors in sexual orientation, the causes and prevention of poverty and crime, the importance of family structure in the likelihood of successful human development and aging, and the list goes on and on. Most important psychosocial factors cannot be manipulated in randomized controlled experiments, so predictive modeling based upon observation research is all that can be accomplished. Unbiased research that is consistently replicated by independent investigators would be especially important in this field, but this may be hard to accomplish with standard predictive modeling methods that can be driven by researchers' biases.

Just as liberal bias is a potential problem in social science, so is conservative bias. For example, economic research on austerity as an effective public policy may have problems not only related to Excel formula errors but also due to the omission of important countries from such an analysis.[23] Yet, given Kahneman's research, errors that favor one's preconceived biases may not be surprising. These biases may not be due to conscious motives or effortful actions, but instead may result from unconscious automatic behaviors. In such a case, it would be very difficult for a researcher with strong ideological convictions to avoid errors or omissions when the results look right because they are in accord with strongly held prior beliefs. But, the effect of this cognitive illusion would be the same as a conscious error even though it is due to completely unconscious factors. A similar conservative bias might have been present in predictive policing research described in a July 19, 2013 Economist.com article. The research claimed to offer support for the hypothesis that predictive policing uncovered more crimes, but the police were not blinded to the model. The police also spent more time in those zones and at those times that were predicted to be high crime probabilities by the model. So, any conclusion that the model targeted policing to cause a drop in crime may be biased by these confounding factors.

Whether there is conservative or liberal bias in social and public policy research or bias related to greed in business and medical applications, faulty predictions and explanations become inputs into a very poor decision-making process. One natural guard against this bias is always to validate models with well-controlled and blinded real world experimental testing. Yet, this only works to show that predictive models cause beneficial effects in the immediate testing period, as environmental changes may occur later in association with spurious and biased predictive factors. So it would be much more efficient always just to build unbiased and reliable models at the outset, which also will be much more likely to be supported in validations. In predictive modeling, the only reasonable guard against this bias appears to be to include as many candidate variables as possible that are based upon diverse and opposing hypotheses about important and causal variables. Yet, in standard predictive modeling methods, this creates another problem related to high-dimensional multicollinear variables. RELR avoids these problems, and thus makes it more likely that the models are not an accident of multicollinearity sampling error. Through its causal learning method based upon outcome score matching, RELR also allows a completely objective method to avoid spurious factors in the predictive model. In the end, RELR will be more likely to avoid those cognitive and data sampling biases that are always a part of observation data. These are the same problems that honest decision makers across business, government, medical, engineering and scientific applications find to be unacceptable.

The neural information theory that is the basis of RELR is actually a work in progress. It is designed to be a theory of high-dimensional Big Data cognitive computations in the brain. This theory is inspired by an attempt to build cognitive machines based upon how neural ensembles might really produce computation and to allow for changes related to new empirical findings. Given the self-improving and self-correcting nature of science, it is likely that at least portions of this theory ultimately will be in need of improvement. So this theory should be viewed as essentially providing a set of questions that need to be asked rather than providing definitive answers in any sense. Yet, the larger message of this book cannot be altered. This is the importance of objective, unbiased, valid modeling predictions and explanations that replicate. If that message is remembered, then the most reliable and valid possible *Calculus Ratiocinator* machines always will be built. With this recommended keen focus on reliability and validity in high dimension data, the era of Big Data could well become known as the era of Big Discovery. And decision makers and scientists who want most likely predictions and explanations confidently will be able to say "Let us calculate".

APPENDIX

Contents

A1. RELR MAXIMUM ENTROPY FORMULATION

Given the equivalence between the maximum entropy and maximum likelihood solutions in logistic regression, RELR can be understood as resulting from the following maximum entropy formulation related to that of Golan, Judge and Perloff.[1] Both maximum likelihood logistic regression and maximum entropy logistic regression are standard textbook methods. There are many excellent sources on these methods and their equivalence, but an especially appealing derivation is one that shows that the maximum entropy derivation avoids the arbitrary guess of a sigmoid S-shaped function in logistic function. Instead, this derivation shows that the sigmoid function results from properties of maximum entropy.[2]

RELR is equivalent in all ways to standard maximum likelihood and maximum entropy logistic regression with the exception that pseudo-observations are used to model error and corresponding error probabilities are estimated. Because of this one fundamental departure, RELR contains constraints and parameters to model error probability in a way consistent with the Extreme Value Type I error in logistic regression.[3] That is, we seek to maximize

$$H(\mathbf{p}, \mathbf{w}) = -\sum_{i=1}^{N}\sum_{j=1}^{C} p_{ij}\ln(p_{ij}) - \sum_{l=1}^{2}\sum_{r=1}^{M}\sum_{j=1}^{2} w_{jlr}\ln(w_{jlr}) \qquad \text{(A1.1)}$$

subject to constraints that include

$$\left(\sum_{i=1}^{N}\sum_{j=1}^{C}(x_{ir}\,y_{ij})\right) + (u_r y_{11r} - u_r y_{12r}) + (u_r y_{22r} - u_r y_{21r})$$

$$= \left(\sum_{i=1}^{N}\sum_{j=1}^{C}(x_{ir}\,p_{ij})\right) + (u_r w_{11r} - u_r w_{12r}) + (u_r w_{22r} - u_r w_{21r}) \quad \text{for}$$

$$r = 1 \text{ to } M, \qquad\qquad\qquad\qquad\qquad\qquad\qquad\qquad\qquad\qquad\qquad \text{(A1.2)}$$

$$\sum_{j=1}^{C} p_{ij} = 1 \quad \text{for } i = 1 \text{ to } N, \qquad\qquad \text{(A1.3)}$$

$$\sum_{j=1}^{2} w_{jlr} = 1 \quad \text{for } l = 1 \text{ to } 2 \text{ and } r = 1 \text{ to } M, \qquad \text{(A1.4)}$$

$$\sum_{i=1}^{N} y_{ij} = \sum_{i=1}^{N} p_{ij} \quad \text{for } j = 1 \text{ to } C - 1, \qquad\qquad \text{(A1.5)}$$

where C is the number of outcome categories or choice alternatives, N is the number of observations and M is the number of data feature constraints. The further right summation in Eqn (A1.1) is the part containing error probability distribution $\mathbf{w}$; the further left summation in Eqn (A1.1) is the part containing the observation probability distribution $\mathbf{p}$. Pseudo-observations are shown as the y_{jlr} in the left-hand side of Eqn (A1.2). They have not appeared explicitly in previous publications because they cancel in the left-hand side of Eqn (A1.2); the left-hand side of these equations also shows real observations as reflected in the y_{ij} terms. In this formulation, $y_{ij} = 1$ if the ith observation yields an outcome that is the jth possible category and 0 otherwise. Also, x_{ir} is the rth independent variable feature associated with the ith observation. In addition to representing noninteractive features, interactions are possible where each interaction value is a product between x_{ir} values which are first standardized or a dummy-coded missing value status indicator for each feature (see below). Each p_{ij} term is always positive and represents the probability that the ith observation yields the jth category as an outcome/choice and w_{jlr} is also always positive and represents the probability of error across pseudo-observations corresponding to the jth category and rth moment and lth sign condition.

As stated at the opening of this section, it is widely known that the maximum likelihood method produces an equivalent solution as this maximum entropy subject to constraints formulation with the objective function shown in Eqn (A1.1). However, standard maximum likelihood logistic regression is an unconstrained optimization problem where the log likelihood that results from the joint probability of all independent events that are modeled is the objective to be maximized. In RELR, this objective contains probabilities of outcome events in real observations or p_{ij} terms and probabilities of error events in pseudo-observations or w_{jlr} in direct analogy with the Eqn (A1.1) when all p_{ij} and w_{jlr} terms are defined identically as the maximum entropy derivation gives as below. This gives the following log likelihood expression in RELR where again the right most summation reflects the pseudo-observations:

$$LL(\mathbf{p}, \mathbf{w}) = \sum_{i=1}^{N} \sum_{j=1}^{C} y(i,j)\ln(p(i,j)) + \sum_{l=1}^{2} \sum_{j=1}^{2} \sum_{r=1}^{M} y(l,j,r)\ln(w(l,j,r))$$

(A1.1a)

As reviewed in the body of this book, the RELR log likelihood that is in Eqn (A1.1a) is often termed *RLL*, and the first summation on the right is called the observation log likelihood or *OLL* and the second summation is called the error log likelihood or *ELL*. As with the constrained maximum entropy expression shown as Eqn (A1.1) and following, the maximum of this *RLL* function can be solved through standard gradient ascent methods.

Other conventions are possible for the coding of pseudo-observations to model error, along with the coding of the direction of error corresponding to the error probability terms w_{jlr}, that give equivalent solutions. Yet, we use an easy to remember convention which is to let the pseudo-observations y_{1lr} corresponding to the $j = 1$ category equal 1 for both positive ($l = 1$) and negative ($l = 2$) error conditions across all $r = 1$ to M features, where positive and negative error directions refer to the sign of the argument of the exponential function in the definition of the error probabilities as shown below for Eqns (A1.9), (A1.10), (A1.12) and (A1.13). Thus, when $l = 1$, a given w_{jlr} term represents the probability of positive error, but when $l = 2$, the w_{jlr} term will represent the probability of negative error. Implicit in this coding is a sign reversal in the interpretation of positive and negative errors, as the $j = 2$ positive and negative errors are actually negations of positive and negative errors at the $j = 1$ condition for all $r = 1$ to M features. So one may consider that the $j = 2$ index has the effect of reversing the sign of u_r, or equivalently that the coefficients of u_r are reversed in sign at the $j = 2$ relative to the $j = 1$ index as explicitly shown in Eqn (A1.2). This makes sense as the sign of the t-value upon which u_r is based would also flip depending upon whether the $j = 1$ or $j = 2$ category is coded as the target value of 1 in the target variable in binary regression or whether these target variable values are coded so that they increase or decrease as j increases across categories in ordinal regression. This is because this t-value is based upon the correlation between the rth independent variable feature and the target variable and there are two t-values—one for positive and one for negative error for each feature (see below).

Because of this sign convention along with the normalization condition of Eqn (A1.4), the basic equivalences $w(1,1,r) = w(2,2,r)$ and $w(1,2,r) = w(2,1,r)$ will apply for all $r = 1$ to M, as can be verified in Eqns (A1.9), (A1.10), (A1.12) and (A1.13) below simply given rules concerning the inversion of the exponential function.

The u_r term is a measure that estimates the expected error for the rth feature. It is defined as

$$u_r = \Omega / \left(r_r / \sqrt{((1 - r_r^2)/N_r)} \right) \quad \text{for}$$

$$r = 1 \text{ to } M, \text{ where } \Omega \text{ is defined as :} \tag{A1.5a}$$

$$\Omega = 2 \sum_{r=1}^{M} |r_r| / \sqrt{((1 - r_r^2)/N_r)} \quad \text{for } r = 1 \text{ to } M, \text{ and where}$$

$$N_r > 2, \quad -1 < r_r < 1, \text{ and where } r_r \neq 0. \tag{A1.5b}$$

where r_r represents the Pearson correlation between the rth feature and the target outcome variable across the N_r nonmissing observations. In the case where r_r equals 1 or -1, the best practice is to adjust it to be approximately 1 or -1 so that division by zero does not occur in Eqn (A1.5a) and (A1.5b).

The part of Eqn (A1.5a) that is the denominator is a t-value that arises from the standard t-test that is performed to determine whether a Pearson correlation between the rth independent variable feature and the target variable is significantly different from zero across the number of nonmissing and independent observations in an independent variable. When this t-value is small, then this error value estimate defined by u_r is large. When this t-value is relatively large, then this u_r error value estimate is relatively small. N_r reflects the number of nonmissing observations across all x_{in} $i = 1$ to N observations where we assume that all observations are independent for each of the $r = 1$ to M features. With a binary outcome variable, r_r is the Pearson product moment correlation between the N_r nonmissing values of the rth feature and the corresponding binary-coded representation of reference versus nonreference membership of the outcome variable, where this binary representation is 1 if the class is the reference class and 0 otherwise. Provided that the sample is stratified so that the numbers of target and nontarget observations are equal in the case of a binary outcome variable, the t-value here will not depend upon assumptions of equal variance in the target and nontarget groups. While such stratification is recommended, in a special case where the sample is not stratified or when there is a mismatch in the number of nonmissing observations in balanced stratified samples in the two groups and where more reliable feature reduction is required, the Welch's t-value is instead employed as in the case of Explicit RELR. This t-value does not assume equal variances in the two groups, and this does give more stable results for such parsimonious variable selection when there is a large difference between the variances of feature values of the two binary groups. Provided that values are missing at random, there will be a strong tendency for roughly equal numbers of missing values in the two binary outcome groups so Welch's t-value will still be approximately similar to Student's t-value. In a case where there are strong structural differences between the two binary groups in terms of missing values with large numbers of missing values only in one group, Welch's t-value that reflects differences between the groups will be much less likely to be a large t-value that reflects an important feature in the feature reduction. In such a case, the important variable will be the dummy-coded missing status indicator which will strongly correlate with the binary outcome variable and will have an equal number of observations in the two binary groups. With ordinal proportional odds logistic regression, r_r is analogous to the Pearson product moment correlation between the nonmissing values of the rth feature and the corresponding values of the ordinal dependent variable. If we assume that ordinal values are expressed as ranks, then Spearman correlations appropriate for ranks and Pearson correlations give equivalent results. Ω is a positively valued scale factor that is the sum of the magnitude of the denominator of these t-values across all M features. This sum is multiplied by 2 because there are two t-values with identical magnitudes but opposite signs for each feature.

The constraints given as Eqn (A1.5) are intercept constraints. These intercept constraints are presented for sake of generality, but in many RELR applications and especially Explicit RELR binary models with perfectly balanced target and nontarget conditions the intercept is dropped from the model. The reason for the separate listing of intercept constraints is because RELR does not attempt to reduce error in intercept weights, as unlike the independent variable features, there are no error probability terms corresponding to these intercept features.

A1.1. Feature Definitions

When nonlinear features are included, the first $M/2$ set of data features in Eqn (A1.2) are from $r = 1$ to $M/2$ and reflect the linear or cubic components that are expected to have the largest magnitude logit coefficients as given in a full model or as defined below through feature reduction. In this case, the second $M/2$ set of data features are from $r = M/2 + 1$ to M and reflect the quadratic or quartic components that are expected to have the largest magnitude logit coefficients as defined similarly. When only linear components are requested, all nonlinear variables are excluded. Nominal input variables are recoded into binary variables for each category condition to be accepted as inputs within this structure, which is different from standard logistic regression which always has one fewer binary categories than total number of categories. Of course, in the case of just two categories, the two binary dummy coded features are perfectly redundant, so one is dropped. In many applications, independent variable features that are perfectly correlated with a previously admitted feature may not be allowed as inputs. Such perfect correlation may be allowed however in special cases, for example where dimension reduction and variable selection is not a primary consideration.

A1.1.1. Linear Constraints

The linear constraints are formed from the original input variables. These constraints are standardized so that each vector $\mathbf{x}_r$ has a mean of 0 and a standard deviation of 1 across all C choice conditions and N observations that gave appropriate nonmissing data. When missing data are present, imputation is performed after this standardization by setting all missing data to 0. This has the effect of adding no information. This is the most benign form of imputation, as it does not try to guess at feature values which is very much unlike multiple imputation. Multiple imputation may lead to false new information if assumptions are not met, so that is why it is not recommended as a preliminary step prior to RELR. The effect of the RELR error model is to ensure that features with more missing values also have more expected error everything else being equal. Interaction input variables are formed by multiplying these standardized variables together to produce the desired interactions. When such interactions are formed, the vector $\mathbf{x}_r$ corresponding to each resulting interaction variable is also standardized and imputed in the same way. Finally, to model missing versus nonmissing patterns of observations in input variables that are correlated to the target variable, new input variables are also formed that are dummy coded to reflect whether observations were missing or not for each previously defined linear component. Through these missing status code variables, structural relationships between missing data and the target variable can also be modeled for each of the components. These missing status code variables are also standardized to a mean of 0 and a standard deviation of 1.

A1.1.2. Quadratic, Cubic, and Quartic Constraints

These constraints are formed by taking elements in each standardized vector $\mathbf{x}_r$ described in Appendix A1.1.1 to the 2nd, 3rd, and 4th power with the exception of the missing code variables. These nonlinear features are themselves restandardized. If the original input variable that formed the linear variable was a binary variable, the nonlinear components will not be independent from linear components and are dropped.

A1.2. Symmetrical Error Probability Constraints on Cross-product Sums

With these feature definitions, two additional sets of linear constraints are now imposed in this maximum entropy formulation. These are constraints on $\mathbf{w}$:

$$\sum_{j=1}^{2} \sum_{r=1}^{M} s_r w_{j1r} - \sum_{r=1}^{M} s_r w_{j2r} = 0 \qquad \text{(A1.6)}$$

$$\sum_{j=1}^{2} \sum_{r=1}^{M} w_{j1r} - \sum_{r=1}^{M} w_{j2r} = 0 \qquad \text{(A1.7)}$$

where s_r is equal to 1 for the linear and cubic group of data constraints and -1 for the quadratic and quartic group of data constraints. Equation (A1.6) forces the sum of the probabilities of error across the linear and cubic components to equal the sum of the probabilities of error across all the quadratic and quartic components. Equation (A1.6) groups together the linear and cubic constraints that tend to correlate and matches them to quadratic and quartic components in likelihood of error. Equation (A1.6) is akin to assuming that there is no inherent bias in the likelihood of positive versus negative error in the linear and cubic components versus the quadratic and quartic components. Equation (A1.7) forces the sum of the probabilities of positive error across all M features to equal the sum of the probabilities of negative error across these same features.

In original publications on RELR, the error model in the far right-hand side of Eqn (A1.1) and all corresponding equations like Eqns (A1.6) and (A1.7) had the outer summation across all C category outcomes. The present formulation gives identical solutions in the case of binary outcomes, but departs in the case of ordinal multinomial outcomes. In the case of ordinal multinomial outcomes, the $j = 1$ to 2 for example in Eqns (A1.6) and (A1.7) categories now simply refer to error modeling categories and do not mean actual ordinal categories. The reason for the change is that practical experience suggested problems with the original approach in that the reference condition was arbitrary and could strongly affect the error model when the direction of the coding was reversed so that the previous highest ordinal category became the lowest ordinal category. This new formulation gives identical solutions without regard to how the reference condition is defined for ordinal dependent variables, where ranks are used to compute the t-values. The presentation of this formulation departs from our previous publications which specifically showed the x feature having $i, j,$ and r indexes for the purposes of ordinal RELR. This is simplified here to just have the i and r indexes because the focus of the book has been on Binary RELR. In a case of Ordinal RELR, obviously the rth value of the feature x at the ith observation should vary across the ordinal category values so it is multiplied by 2 for $j = 2$ and by 3 for $j = 3$ etc. and 0 for the reference. The ordinal RELR formulation here is close to how proportional odds ordinal logistic regression is shown in standard logistic regression texts like Hosmer and Lemeshow. Yet, Hosmer and Lemeshow admit that the meaning of the β coefficients is somewhat imprecise. So the ordinal category level needs to be considered also in ordinal probability definitions and this is shown below.[4]

A1.3. Form of Solutions

This maximum entropy subject to constraints solution is found through the standard constrained optimization method which implies setting up a Lagrangian multiplier system and then solving for the optimal solution through a gradient ascent method for example.[5] The probability components that are of interest in the solutions have the form

$$p_{ij} = \exp\left(\alpha_j + \sum_{r=1}^{M}(C-j)\beta_r x_{ir}\right) \Big/ \left(1 + \sum_{k=1}^{C-1}\exp\left(\alpha_k + \sum_{r=1}^{M}(C-k)\beta_r x_{ir}\right)\right) \quad \text{for}$$

$$i = 1 \text{ to } N \text{ and } j = 1 \text{ to } C-1,$$

$$\text{(A1.8)}$$

$$w_{j1r} = \exp(\beta_r u_r + \lambda + s_r \tau)/(1 + \exp(\beta_r u_r + \lambda + s_r \tau)) \quad \text{for}$$

$$r = 1 \text{ to } M \text{ and } j = 1, \tag{A1.9}$$

$$w_{j2r} = \exp(-\beta_r u_r - \lambda - s_r \tau)/(1 + \exp(-\beta_r u_r - \lambda - s_r \tau)) \quad \text{for}$$

$$r = 1 \text{ to } M \text{ and } j = 1. \tag{A1.10}$$

$$p_{ij} = 1\Big/\left(1 + \sum_{k=1}^{C-1}\exp\left(\alpha_k + \sum_{r=1}^{M}(C-k)\beta_r x_{ir}\right)\right) \quad \text{for}$$

$$i = 1 \text{ to } N \text{ and } j = C, \text{which is the reference condition,} \quad \text{(A1.11)}$$

$$w_{j1r} = 1/(1 + \exp(\beta_r u_r + \lambda + s_r \tau)) \quad \text{for } r = 1 \text{ to } M \text{ and } j = 2,$$

$$\text{(A1.12)}$$

$$w_{j2r} = 1/(1 + \exp(-\beta_r u_r - \lambda - s_r \tau)) \quad \text{for } r = 1 \text{ to } M \text{ and } j = 2.$$

$$\text{(A1.13)}$$

Note that α terms that are indexed as $k = 1$ to $C-1$ and $j = 1$ to $C-1$ are intercept weights which are defined in Eqn (2.5a) and (2.5b) for binary logistic regression without the index as there is only one intercept but there can be more than one intercept in ordinal regression. As noted above, obviously the feature x should be also indexed by the category indicator j or k in the case of Ordinal RELR so that it is multiplied by that category

number or a value of 0 for the reference number in the probability definitions. This multiplication factor is now substituted into these probability
definition equations to give a correct definition for Ordinal RELR that is
equivalent to explicitly having each x feature vary across ordinal categories.
The critical assumption in RELR as in standard maximum likelihood logistic regression is that all observations are independent. In the case of any
offset weight as is used in Implicit RELR's offset regression, Eqns (A1.8)
and (A1.11) can have the argument to the exponential modified to reflect
additive terms O_{ij}, $j = 1$ to $C - 1$ and $i = 1$ to N, O_{ik}, $k = 1$ to $C - 1$ and
$i = 1$ to N to capture this offset weight multiplied by the feature $x(i,r)$
for any ith individual and rth feature which is offset. In general, these
equations are

$$p_{ij} = \exp\left(O_{ij} + \alpha_j + \sum_{r=1}^{M}(C-j)\beta_r x_{ir}\right) \Big/ \left(1 + \sum_{k=1}^{C-1}\exp\left(O_{ik} + \alpha_j + \sum_{r=1}^{M}(C-k)\beta_r x_{ir}\right)\right)$$

for $i = 1$ to N and $j = 1$ to $C - 1$,

(A1.8a)

$$p_{ij} = 1\Big/\left(1 + \sum_{k=1}^{C-1}\exp\left(O_{ik} + \alpha_j + \sum_{r=1}^{M}(C-k)\beta_r x_{ir}\right)\right) \quad \text{for } i$$

$$= 1 \text{ to } N \text{ and } j = C,$$

(A1.11a)

where O_{ij} and O_{ik} are defined as

$$O_{ih} = (C - h)\beta_o(r)x(i, r) \quad \text{for } h = j = 1 \text{ to } C - 1 \text{ or } h = k$$

$$= 1 \text{ to } C - 1,$$ (A1.11b)

for the rth feature which is being used as the offset effect. In this special case
of the Implicit RELR offset regression used in RELR's outcome probability
matched samples, only one feature is typically used as an offset so $\beta_o(r)$ in
Eqn (A1.11b) only would be nonzero for this one offset feature as $\beta_o(r)$
would be zero for all other features and there also would be a zero intercept
typically. In addition, in this case of the Implicit RELR offset regression
used in RELR's outcome score matching, the regression coefficient $\beta(r)$ in
Eqns (A1.8a) and (A1.11a) is zero for the offset feature, as the regression
effect defined by Eqn (A1.11b) is used instead which uses $\beta_o(r)$ for this offset
feature. In contrast, all regression coefficients can be updated in RELR's KL

updating, and this derivation is shown in Appendix A4 as it uses slightly different symbols that reflect prior and update weights.

A1.4. RELR and Feature Reduction

With high-dimensional data, the purpose of feature reduction is to reduce the dimensionality of the problem into a manageable number of independent variable features. The basic idea is to capture the subset of features that would have the largest magnitude logit coefficients in the full model, while excluding those features with smaller magnitude logit coefficients. In this way, we can build a model with a much smaller set of features than with the entire set and achieve a model that still includes those features with the largest weight in the ultimate solution after feature selection as if a feature selection model had been built with all features included originally. This is important, as it is very easy to get models with tens of thousands of features when higher level interactions are included, even when the original number of variables was less than 100. This features process arises from a set of relationships in the preceding equations. For example, from Eqns (A1.9), (A1.10), (A1.12) and (A1.13), we have

$$\ln(\omega_{11r}/\omega_{12r}) - \ln(\omega_{21r}/\omega_{22r}) = 2\beta_r u_r + 2\lambda + 2s_r\tau \quad r = 1 \text{ to } M. \tag{A1.14}$$

The right-hand side of Eqn (A1.14) is interpreted as the difference in positive and negative errors for the rth feature. This error ε_r is directly related to the feature error u_r:

$$\varepsilon_r = 2\beta_r u_r + 2\lambda + 2s_r\tau \quad \text{for } r = 1 \text{ to } M, \tag{A1.15}$$

and is assumed to be Extreme Value Type I distributed because it arises as the difference between positive and negative error variables that have probabilities that are distributed according to the logistic distribution; a difference between logistic distributed variables was reviewed in Chapter 2 to generate such extreme value distributed error. Writing u_r in terms of Ω/t_r and rearranging terms gives

$$(t_r/\Omega)(\varepsilon_r/2 - \lambda - s_r\tau) = \beta_r \quad \text{for } r = 1 \text{ to } M. \tag{A1.16}$$

Therefore, we know that the following relationship will hold when each ε_r can be assumed to have a negligible value so it can be assumed to close to zero:

$$(t_r)(-\lambda - s_r\tau)/\Omega \approx \beta_r \quad \text{for } r = 1 \text{ to } M. \tag{A1.17}$$

Hence, we know that the value of each logit coefficient β_r will be approximately proportional to t_r across all linear and cubic features, but with a different proportionality across all quadratic and quartic features, when in all cases ε_r is close to zero. This is because we also know that the expression $-\lambda - s_r\tau$ corresponding to all linear and cubic variables will be equal across all $r = 1$ to $M/2$ components and that corresponding to all quadratic and quartic variables also will be equal across all $r = M/2 + 1$ to M features for each of the jth conditions. This follows from the definition of s_r in Section 1.2. Therefore, we use this relationship to select the linear and/or cubic features with the largest expected logit coefficient magnitudes simply in terms of the magnitude of t_r. Likewise, we select the largest expected logit coefficient magnitudes for all quadratic and/or quartic features simply in terms of this same magnitude. In other words, this allows us to perform feature reduction in a way that will not influence the ultimate Implicit or Explicit RELR feature selection provided that our feature reduction set is large enough, as we will simply discard the noisy features through feature reduction that would be unimportant anyway after feature selection.

In fact, we expect that this relationship in Eqn (A1.17) to be a very good approximation when Ω gets large in comparison to the magnitude of t_r. This is due to the linear relationship in Eqn (A1.2) which suggests that if we substitute $u_r = \Omega/t_r$, then as Ω gets larger in comparison to the magnitude of t_r, w_{11r} and w_{12r} must become closer to being equal. At very large values of $\Omega/|t_r|$, w_{11r} and w_{12r} would be substantially equal, but this would require that ε_r is also close to zero which is exactly what we assume. Then again, this relationship in Eqn (A1.17) could be a poor approximation for data which have few features with significantly large t-value magnitudes. This is because there would be small values of the ratio $\Omega/|t_r|$ in which case the error ε_r is not necessarily close to zero. However, in such data, the features with the strongest relationship to the target variable would not be an issue as these are the few features with significantly large t-value magnitudes. If these were causal features, this would be akin to assuming that cause is not possible without some degree of correlation. Therefore, in general, we use the magnitude of t_r to select the features with the largest expected logit coefficient magnitudes whether or not the actual estimated error ε_r turns out to be close to zero. With many features with relatively large t-values, empirical results do consistently verify this reliably strong relationship between t_r and β_r that allows feature reduction to get highest magnitude logit coefficients simply on the basis of t_r magnitudes. Yet, because this correspondence between t_r and β_r tends to deviate from exact proportionality more and more as there are fewer features with relatively large magnitude t-values, it is not often seen in Explicit RELR solutions that result from parameter deletion.

RELR's lack of necessary rigid proportionality between t_r and β_r differentiates it from Naïve Bayes. For example, given standardized variables with no missing data, the Gaussian Naïve Bayes formulation is rigidly tied to exact proportionality between a measure of t_r and β_r that does not allow for interactions.[6] In contrast, RELR is well suited for high-dimensional data and interactions, as RELR's t-value screening is easily processed in parallel.

A2. DERIVATION OF RELR LOGIT FROM ERRORS-IN-VARIABLES CONSIDERATIONS

In the RELR method, parameters that capture the effect of error are directly estimated. Thus, the RELR logit can be shown to result from error-in-variables considerations concerning the effect of the joint probability of all estimated error. This results from the independence assumptions concerning the outcome and error events. That is, given definitions of the outcome probabilities and error probabilities shown in Appendix A1, the joint probability of a binary outcome event at the ith observation and $j = 1$ category and all error events across the $l = 1$ to 2 positive and negative error-in-variables conditions across the $r = 1$ to M features takes the form

$$p(i, 1)w(1, 1, 1)...w(1, 2, M) = p(i, 1) \prod_{r=1}^{M} \prod_{l=1}^{2} w(1, l, r)$$

$$\text{for } i = 1 \text{ to } N. \tag{A2.1}$$

And similarly for $j = 2$ this joint probability becomes

$$p(i, 2)w(2, 1, 1)...w(2, 2, M) = p(i, 2) \prod_{r=1}^{M} \prod_{l=1}^{2} w(2, l, r)$$

$$\text{for } i = 1 \text{ to } N. \tag{A2.2}$$

Any consideration of the ratio of the joint probability defined in Eqn (A2.1) relative to that defined in Eqn (A2.2) may at first seem unwieldy. But because $w(1,1,r) = w(2,2,r)$ and $w(1,2,r) = w(2,1,r)$ will apply for all $r = 1$ to

M as noted in Appendix A1, all of these terms describing the probability of error simply vanish. So when we take the log odds of the ratio of these joint probabilities, we simply get the log odds of the binary outcome described by $p(i,1)$ because $p(i,2) = 1 - p(i,1)$. So, the logit of the outcome event described by $p(i,1)$ can be written as

$$\text{logit}(i, 1) = \sum_{r=1}^{M} \sum_{l=1}^{2} \varepsilon(l, 1, r) + \alpha + \sum_{r=1}^{M} \beta(r) x(i, r) \quad \text{for } i = 1 \text{ to } N,$$

(A2.3)

which represents the positive and negative errors across the $r = 1$ to M features, but it is recognized that these far left-most summations involving the error terms cancel out to zero. So even though this has the same form as standard logistic regression where we can ignore the error parameters in a scoring model because the error sums to zero, the errors in the logit have been estimated through an error-in-variables method that accounts for positive and negative errors at each of the $r = 1$ to M features. Thus, the estimates $\beta(r)$ $r = 1$ to M will be adjusted for this error-in-variables at each of the $r = 1$ to M independent variable features. Because of this, these regression estimates and this logit almost always will differ from that which is found in standard logistic regression due to this error-in-variables adjustment. The only major exception will be when only one feature is estimated in the model, as RELR is equivalent to standard logistic regression in that special case.

A3. METHODOLOGY FOR PEW 2004 ELECTION WEEKEND MODEL STUDY

Data were obtained from the Pew Research Center's 2004 Election Weekend survey using observations that indicated Bush or Kerry as their vote. In a first "smaller training sample" model, the training sample consisted of a simple random sample of 8% or 188 observations. The remainder of the overall sample defined the validation sample. In a second "larger training sample" model, the training sample consisted of a simple random sample of roughly 50% or 1180; the remainder of this sample defined the validation sample. This sampling method was the default sampling method in the SAS Enterprise Miner 5.2 software that was also used to build the models. The target variable was Presidential Election Choice (Bush versus Kerry). Kerry was the target condition. The 2004 election was very close, as roughly 50% of the respondents opted for Kerry in all of these subsamples. Hence, these are extremely balanced samples, but not quite perfectly balanced.

There were 11 interval variables and 44 nominal input variables originally, but the nominal variables were recoded into binary variables for input into RELR. RELR also produced new input variables from these original input variables corresponding to two-way interactions, polynomial terms, and missing data. Over 2500 variables resulted in total. Both Implicit and Explicit feature selection models were run. 176 variables were selected in the Implicit RELR feature selection from the larger training sample, whereas 24 variables were selected in the Implicit RELR smaller training sample model. These identical variables were input into Penalized Logistic Regression (PLR) to have an "apples-to-apples" comparison to Implicit RELR. For this same "apples-to-apples" reason, the same intercept estimation procedure was used with RELR models and PLR, including the use of thresholds that minimized bias in classification with corresponding intercept correction. The λ value that is the penalty parameter in these PLR models was systematically varied between 0.5 and 150; values above 150 had floating point error. In both samples, the $\lambda = 1$ was either associated with the best validation misclassification rate as in the small sample, or very close to the best. However, we report the results based upon $\lambda = 1$ for both samples here because this is in the range that is normally employed for λ for PLR.

Bush versus Kerry models were also run within Enterprise Miner 5.2 software using the Support Vector Machine, Partial Least Squares, Decision Tree, Logistic Regression and Neural Network methods. SVM and PLS were beta versions in EM 5.2; reported error measures were computed independently from EM for these methods and RELR. Defaults were employed in all cases except Logistic Regression where two-way interactions (between two independent variable features) and polynomial terms up to the 3rd degree were specified and Support Vector Machines where the polynomial kernel function was requested. Also, Stepwise Selection was performed with the Logistic Regression in the Pew "smaller sample", but no selection was performed with the larger sample due to the long time that it took to run the stepwise process in this sample. In addition, the Imputation Node within Enterprise Miner was employed with the Regression and Neural Network methods and was run using its default parameters. The identical variables and samples were employed as inputs in all cases. Like most political polling data sets, there was a wide range in the correlation between the original input variables that went from roughly -0.6 to about 0.81. These correlation magnitudes were over 0.9 for many of the interactions and nonlinear variables produced by RELR, so this data set clearly exhibited multicollinearity. In addition, there was significant correlation to the target variable in a number of variables; the largest correlations were in the $0.7–0.8$ range.

A4. DERIVATION OF POSTERIOR PROBABILITIES IN RELR'S SEQUENTIAL ONLINE LEARNING

This appendix reviews the basic derivation of RELR's posterior probabilities as related to minimal KL divergence/maximum likelihood, and then also reviews some nuances that occur with time-varying data like survival and seasonality effects. Finally, RELR's sequential online learning method is graphically depicted in Fig. A4.1 below.

To become probabilities, the posterior probabilities need to be normalized in Eqn (3.3a) and (3.3b) so that they sum to unity across the $j = 1$ to C category conditions for each ith observation for $i = 1$ to N. This section of the appendix will show this for the simple case with binary outcomes so $j = 1$ to 2:

Given that $\exp(x)/(1 + \exp(x)) = \exp(x/2)/(\exp(-x/2) + \exp(x/2))$, and normalizing to probabilities causes the posterior probabilities defined in Eqn (3.3a) and (3.3b) to become

$p(i, j = 1)$

$$= \frac{q(i,j = 1)e^{\left(\Delta\alpha + \sum_{r=1}^{M} \Delta\beta(r)x(i,r)\right)/2}}{((1 - q(i,j = 1))e^{\left(-\Delta\alpha - \sum_{r=1}^{m} \Delta\beta(r)x(i,r)\right)/2} + q(i,j = 1)e^{\left(\Delta\alpha + \sum_{r=1}^{m} \Delta\beta(r)x(i,r)\right)/2}}$$

$$(A4.1)$$

$p(i, j = 2)$

$$= \frac{(1 - q(i,j = 1))e^{\left(-\Delta\alpha - \sum_{r=1}^{M} \Delta\beta(r)x(i,r)\right)/2}}{((1 - q(i,j = 1))e^{\left(-\Delta\alpha - \sum_{r=1}^{m} \Delta\beta(r)x(i,r)\right)/2} + q(i,j = 1)e^{\left(\Delta\alpha + \sum_{r=1}^{m} \Delta\beta(r)x(i,r)\right)/2}}$$

$$(A4.2)$$

for $i = 1$ to N where the prior probability $q(i,j = 2) = 1 - q(i,j = 1)$ for all $i = 1$ to N observations. Kullback–Leibler or KL updating is a Bayesian online learning process with the old posterior solution becoming the new prior distribution for each new observation episode when we view the prior

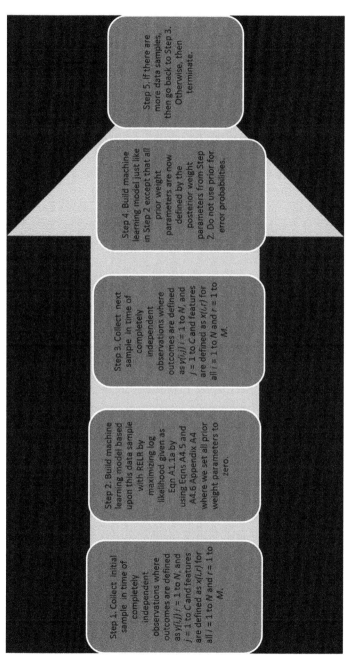

Figure A4.1 RELR's sequential online learning method.

probabilities as being built from prior regression weights from a previous observation episode where

$$q(i,j = 1) = \frac{e^{\left(\alpha_q + \sum_{r=1}^{M} \beta_q(r)x(i,r)\right)/2}}{e^{\left(-\alpha_q - \sum_{r=1}^{m} \beta_q(r)x(i,r)\right)/2} + e^{\left(\alpha_q + \sum_{r=1}^{m} \beta_q(r)x(i,r)\right)/2}} \qquad (A4.3)$$

$$q(i,j = 2) = \frac{e^{\left(-\alpha_q - \sum_{r=1}^{M} \beta_q(r)x(i,r)\right)/2}}{e^{\left(-\alpha_q - \sum_{r=1}^{m} \beta_q(r)x(i,r)\right)/2} + e^{\left(\alpha_q + \sum_{r=1}^{m} \beta_q(r)x(i,r)\right)/2}} \qquad (A4.4)$$

for $i = 1$ to N and where α_q and β_q reflect prior parameters based upon prior observations. So if we simply substitute these expressions for $q(i,j = 1)$ and $1 - q(i,j = 1) = q(i,j = 2)$ into Eqns (A4.1) and (A4.2) and remember again that $\exp(x)/(1 + \exp(x)) = \exp(x/2)/(\exp(-x/2) + \exp(x/2))$ and $1/(1 + \exp(x)) = \exp(-x/2)/(\exp(-x/2) + \exp(x/2))$, we get

$$p(i,j = 1) = \frac{e^{\left(\Delta\alpha + \sum_{r=1}^{M} \Delta\beta(r)x(i,r)\right) + \left(\alpha_q + \sum_{r=1}^{M} \beta_q(r)x(i,r)\right)}}{1 + e^{\left(\Delta\alpha + \sum_{r=1}^{M} \Delta\beta(r)x(i,r)\right) + \left(\alpha_q + \sum_{r=1}^{M} \beta_q(r)x(i,r)\right)}} \qquad (A4.5)$$

$$p(i,j = 2) = \frac{1}{1 + e^{\left(\Delta\alpha + \sum_{r=1}^{M} \Delta\beta(r)x(i,r)\right) + \left(\alpha_q + \sum_{r=1}^{M} \beta_q(r)x(i,r)\right)}} \qquad (A4.6)$$

for $i = 1$ to N. This can be solved using maximum likelihood estimation exactly as previous probability definitions given in Chapter 2, where we now only now need to remember that the weights α_q and $\beta_q(r)$, $r = 1$ to M are the prior weights used to construct the prior probability distribution $\mathbf{q}$ typically based upon historical empirical observations so they are fixed and cannot change in the maximum likelihood estimation. Note that the error probability distribution $\mathbf{w}$ in Eqn (A1.1) is not affected by prior probabilities in this RELR KL updating method. Equations (A4.5) and (A4.6) are similar to the form of the observation probability definitions for binary standard logistic regression and RELR reviewed in Chapter 2 (Eqn (2.5a) and (2.5b)). Yet, we recognize that α and $\beta(r)$, $r = 1$ to M were the posterior weights in Eqn (2.5a) and (2.5b). And these weights are just the sum of the weights that are now prior weights α_q and $\beta_q(r)$, $r = 1$ to M and the update weights $\Delta\alpha$ and $\Delta\beta(r)$, $r = 1$ to M given in Eqns (A4.5) and (A4.6). Note that for RELR's ordinal regression, Eqns (A4.5) and (A4.6) need to be modified in analogy with Eqns (A1.8) and (A1.11) to accommodate more than two categories C, but this is very straightforward.

In some cases it is an advantage not to update one or more parameters with sequential online learning. For example in survival analysis or with seasonal variation, some parameters can be time varying with a value that is not then updated through sequential online learning. These parameters can have an index $t = 1$ to T to reflect the time-varying properties.[7] As one example, the intercept for each period of time can be interpreted as the expected or baseline hazard rate across individuals when binary logistic regression is applied to data that are binary coded and censored in a way that is appropriate for survival data.[8] But RELR must model successive sequences of independent observation samples drawn from these data just like it handles all other sequential sampling for online learning. To allow the intercept to vary over time and have a unique value for each successive time period without being updated, each prior intercept weight $\alpha_q(t)$ can be set to zero for these time periods. So the update weight or $\Delta\alpha(t)$ then is not influenced by those prior periods and this update weight $\Delta\alpha(t)$ $t = 1$ to T then can be interpreted as the baseline hazard rate in each select period of time in this application. Select $\Delta\beta(r)$ values similarly could become temporally dependent parameters $\Delta\beta(r,t)$, $t = 1$ to T simply by setting their prior weight parameters $\beta_q(r,t)$ to zero for $t = 1$ to T and now allowing these parameters to change over time.[9] In the case of a perfectly balanced binary stratified sample and zero intercept, the intercept can be corrected using the usual procedure.[10] In this case, the corrected intercept is the value that is stored as the time-varying intercept $\Delta\alpha(t)$ $t = 1$ to T.

In some applications, it can be expected that there would be time variation across observations in terms of when an observation occurs within a given cross-section or cohort sample. The precise time of the observation then can become a covariate feature that can have associated $\beta_q(r,t)$ and $\Delta\beta(r,t)$ parameters modeled, or which can have interaction effects with other covariates modeled. So a baseline hazard rate or seasonal estimate thus can be adjusted by temporal covariate features across individual observations in this way, and through the interaction of other covariate features with this temporal covariate feature.

A5. CHAIN RULE DERIVATION OF EXPLICIT RELR FEATURE IMPORTANCE

The Explicit RELR feature importance measure defined in Eqn (3.5b) is easily computed based upon an assumption that the sample is large enough so that RELR's Wald-estimated χ^2 values are good approximations to the more reliable likelihood ratio estimates of χ^2. In standard logistic regression this assumption cannot be made until the sample is fairly large, but in RELR this is a very reasonable assumption with very small sample sizes. Given the Wald approximation to χ^2 in logistic regression, the χ^2 value for the rth feature can be defined in terms of the regression coefficient β and se or standard error of that coefficient as

$$\chi_r^2 = (\beta_r/se_r)^2 \quad \text{for } r = 1 \text{ to } M, \tag{A5.1}$$

or writing this in terms of stability s which is the inverse standard error where $s = 1/se$

$$\chi_r^2 = (\beta_r s_r)^2 \quad \text{for } r = 1 \text{ to } M. \tag{A5.2}$$

But

$$d\chi_r^2/d(s_r\beta_r) = 2\chi_r \quad \text{for } r = 1 \text{ to } M, \tag{A5.3}$$

and

$$d(s_r\beta_r)/ds_r = \beta_r \quad \text{for } r = 1 \text{ to } M. \tag{A5.4}$$

So from the chain rule

$$d\chi_r^2/ds_r = (d\chi_r^2/ds_r\beta_r)(d(s_r\beta_r)/ds_r) \quad \text{for } r = 1 \text{ to } M, \tag{A5.5}$$

or

$$d\chi_r^2/ds_r = 2\chi_r\beta_r \quad \text{for } r = 1 \text{ to } M. \tag{A5.6}$$

So the magnitude of this derivative can be defined as

$$\left| d\chi_r^2/ds_r \right| = \left| 2\chi_r\beta_r \right| \quad \text{for } r = 1 \text{ to } M. \tag{A5.7}$$

This is the Explicit RELR feature importance measure that is defined by Eqn (3.5b) in Chapter 3.

A6. FURTHER DETAILS ON THE EXPLICIT RELR LOW BIRTH WEIGHT MODEL IN CHAPTER 3

Table A6.1 shows key parameters in the Explicit RELR feature selection at specific iterations in the trajectory shown in Fig. 3.4(a) and (b), along with the univariate t-values across features. In the 1st iteration, t-values and β coefficients are roughly proportional. Because χ is roughly similar across features here in the 1st iteration, the feature importance measure $\left| d\chi^2/ds \right|$ which is based upon the product of χ and β is also roughly proportional

Table A6.1 Key parameters from Explicit RELR feature selection trajectory shown in Fig. 3.4(a) and (b) at the 1st, 7th, 8th, and 9th iterations. This also shows t-values across features in the middle of the table. Note that u is the expected feature error defined as twice the sum of all t-value magnitudes across all remaining features divided by the t-value for a feature, and the Error Coefficient is the coefficient that results from the symmetrical error constraint which forces an equal probability of positive and negative errors. The error probabilities are shown here across features as $w(pos)$ and $w(neg)$

1st iteration	Birth	Smoke	Whiterace	Blackrace	Otherrace	Age	LWT	Prevlowtotal	Lastlow
u	32.06	−123.85	13.17	−43.85	−15.92	39.87	−293.71	7.43	7.35
χ	2.14	2.40	2.25	2.37	2.40	2.27	2.40	2.76	2.69
β	0.17	−0.05	0.43	−0.13	−0.35	0.14	−0.02	0.79	0.79
se	0.08	0.02	0.17	0.05	0.15	0.06	0.01	0.29	0.29
$|d(\chi^2)/ds|$	0.71	0.22	2.15	0.61	1.68	0.63	0.09	4.38	4.24
$u(pos)$	0.58	0.49	0.50	0.50	0.51	0.53	0.49	0.43	0.46
$u(neg)$	0.42	0.51	0.50	0.50	0.49	0.47	0.51	0.57	0.54
Error coefficient = −5.63									
$t(univariate)$	1.03	−0.27	2.51	−0.75	−2.08	0.83	−0.11	4.45	4.50

7th Iteration	Whiterace	Smoke	Lastlow	Prevlowtotal
u	9.13	5.15	5.09	4.02
χ	2.71	2.50	2.18	1.87
β	0.61	1.04	1.01	0.92
se	0.23	0.42	0.46	0.49
$|d(\chi^2)/ds|$	3.32	5.22	4.40	3.43
$u(pos)$	0.44	0.50	0.56	0.48
$u(neg)$	0.56	0.50	0.44	0.52
Error coefficient = −5.37				

8th Iteration	Prevlowtotal	Lastlow
u	4.02	3.98
χ	1.87	1.53
β	0.92	0.88
se	0.49	0.58
$|d(\chi^2)/ds|$	3.43	2.70
$u(pos)$	0.48	0.52
$u(neg)$	0.52	0.48
Error coefficient = −3.60		

9th Iteration	Prevlowtotal
u	2.00
χ	4.55
β	1.84
se	0.40
$|d(\chi^2)/ds|$	16.71
$u(pos)$	0.50
$u(neg)$	0.50
Error coefficient = −3.67	

with the magnitude of t across features. Yet, these relationships break down in later iterations, where Explicit RELR is no longer governed by these univariate relationships. Table A6.2 shows these same parameters in the identical iterations in a sample that merged this test sample of 56 observations with the validation sample of 60 observations. Notice now that these univariate relationships are not quite as obvious at the 1st iteration, as $|d\chi^2/ds|$ is no longer roughly proportional with the magnitude of t across features and χ is now more variable showing a greater range in Table A6.2 compared with Table A6.1. So this is a good example of how Explicit RELR feature selection behaves like a feature selection based upon univariate measure of importance when samples are very small and/or when multicollinear features are present in more important features especially in higher dimensions. Yet, with larger samples or lower dimensions, Explicit RELR is no longer dropping features based upon univariate measures of importance.

RELR always needs to assume that a large enough sample is collected so that t-values are reliable. This assumption is not reasonable for the smallest magnitude t-values in the very small training sample of only 56 observations here, but is it is reasonable for the larger magnitude t-values here. Because all the χ^2 values even for very small β values are roughly the same in higher dimension and/or small sample models for comparable polynomial effects whether odd or even, the χ^2 values all have roughly the same p value which in this case with one degree of freedom would $p < 0.05$. Clearly there is more variability in the β coefficients with small t-values than would be predicted by that inference as reflected by the fact that some even reverse in sign at the larger sample. Yet, RELR is still fairly robust when it has features with larger magnitude t-values as in this example. Still, the maximal RLL solution in the smaller training sample returns a regression coefficient of 1.84 for the PREVLOWTOTAL feature which is biased high in magnitude compared with the larger sample's solution which had a regression coefficient of 1.52, although it is within the 95% confidence interval of $+/- 2se = +/- 0.8$. Such a one feature RELR model without an intercept will be exactly the same model as standard logistic regression.

Hosmer and Lemeshow report on a model developed with a standard logistic regression method comparable with Stepwise Selection called Best Subsets Logistic Regression. This model is based upon this same Low Birth Weight data used in the current Explicit RELR model, and includes independent variables from a similar cross-sectional sample to predict the

Table A6.2 Key parameters from a full sample composed of both training and validation observations which totaled 116 in this case from the Explicit RELR feature selection trajectory at the 1st, 7th, 8th, and 9th iterations. In all ways, these parameters are defined similarly as in Table A6.1 except that this is based upon a larger sample composed of both training ($n = 56$) used in Table A6.1 and the previous validation sample ($n = 60$) used in Table 3.1

1st Iteration	Birth	Smoke	Whiterace	Blackrace	Otherrace	Age	LWT	Prevlowtotal	Lastlow
u	128.70	40.05	14.10	−28.57	−21.33	−212.21	−36.69	7.05	7.11
χ	2.06	1.99	2.16	2.24	2.02	2.25	2.42	2.88	2.47
β	0.03	0.10	0.29	−0.16	−0.19	−0.02	−0.13	0.71	0.65
se	0.02	0.05	0.13	0.07	0.10	0.01	0.05	0.25	0.26
$|d(\chi^2)/ds|$	0.14	0.42	1.24	0.71	0.79	0.10	0.64	4.08	3.21
$u(pos)$	0.54	0.57	0.60	0.49	0.58	0.48	0.41	0.37	0.46
$u(neg)$	0.46	0.43	0.40	0.51	0.42	0.52	0.59	0.63	0.54
Error coefficient = −4.46									
t(univariate)	0.29	0.94	2.66	−1.31	−1.76	−0.18	−1.02	5.31	5.27

7th Iteration	Whiterace	Prevlowtotal	Lastlow	8th Iteration	Prevlowtotal	Lastlow	9th Iteration	Prevlowtotal
u	9.97	4.98	5.02	u	4.02	3.98	u	2.00
χ	2.47	2.41	1.75	χ	1.99	1.31	χ	5.48
β	0.39	0.79	0.70	β	0.79	0.70	β	1.52
se	0.16	0.33	0.40	se	0.40	0.53	se	0.28
$|d(\chi^2)/ds|$	1.93	3.83	2.46	$|d(\chi^2)/ds|$	3.16	1.83	$|d(\chi^2)/ds|$	16.65
$u(pos)$	0.47	0.46	0.57	$u(pos)$	0.46	0.54	$u(pos)$	0.50
$u(neg)$	0.53	0.54	0.43	$u(neg)$	0.54	0.46	$u(neg)$	0.50
Error coefficient = −3.79				Error coefficient = −2.98			Error coefficient = −3.04	

next cross-sectional sample in transition model. AGE, SMOKE, and LWT were selected as variables, along with a variable that they call PREVLOW which is identical to the LASTLOW variable here. Hosmer and Lemeshow calculated that their regression coefficient of 3.415 translates into an odds ratio of 30. The interpretation is that the odds are roughly 30 times greater of a low birth weight pregnancy in a woman whose previous pregnancy had a low birth weight compared with one whose previous pregnancy was not a low birth weight. They suggest that this does seem unrealistically large. In the present Explicit RELR model, there are roughly 10 times greater odds (95% confidence interval for odds ratio = 2.56–39.0) that there will be a low birth weight pregnancy in a woman with only one previous low birth weight pregnancy when the standard regression coefficient is considered and the intercept is readjusted to correct for the balancing (see next chapter). Yet, RELR does show substantially higher odds ratio of 516 or greater when two or more previous low weight births have occurred compared with no previous low weight births. So while the Explicit RELR model is much more parsimonious than the Hosmer and Lemeshow model as it only selects one independent variable and its odd ratio of 10 for the condition of whether only one previous low birth weight relative to none may be reasonable, it may be similarly biased in terms of the magnitude of the regression coefficient that predicts low birth weight based upon previous history based upon have had two or more low birth weight pregnancies. Such bias is expected in RELR's solution when only one feature is selected, as this is the same solution that standard logistic regression would give. So, the only reason that RELR had lower bias than Hosmer and Lemeshow here is that its feature selection model had only one feature, whereas the Hosmer and Lemeshow Stepwise method selected multiple features which is known to increase bias. Interestingly, the case–control matched sample quasi-experiment reviewed in Chapter 4 shows that the odds ratio of a low birth weight pregnancy for a woman having had any low birth weight pregnancy relative to none is only about 7, which can be interpreted as showing less bias. So, even when Explicit RELR's data mining gives identical values as standard logistic regression would give with only one feature being selected, the RELR outcome score matching causal learning method still may produce models with much lower bias in regression coefficients. This does accord with the proposal that RELR's outcome score matching may be considered to be a better assessment of expected causal effects than Explicit RELR's data mining.

A7. ZERO INTERCEPTS IN PERFECTLY BALANCED STRATIFIED SAMPLES

The model shown in Fig. 3.4(a) and (b) is different in one fundamental sense from all Explicit and Implicit RELR models that have been reported previously here or anywhere else. This is that the intercept has been set to zero in a completely balanced stratified sample that included half target responses and half nontarget responses (Low Birth Weight and Non-Low Birth Weight pregnancies). It turns out that the removal of the intercept in a completely balanced sample has implications in terms of selecting features that are less prone to bias and convergence failures compared with a model which keeps the intercept in the selection set. These problems that result when an intercept is included are shown in Fig. 4.1. This Explicit RELR model is based upon the identical Low Birth Weight training sample employed in Fig. 3.4(a) and (b). The only difference is that the intercept is now included in the model in Fig. A7.1(b), whereas it was excluded in the model in Fig. 3.4(b).

In Fig. A7.1(b), it is seen that the intercept grows to be unrealistically large and biased in magnitude in the 7th iteration and the model actually failed to converge in the critical 8th and 9th iterations. Often, there will be convergence when the intercept grows to a large value, but the problem is that regression coefficients in selected features are also biased to be unrealistically large in magnitude. In the development of RELR, it was originally thought that it would be optimal to keep the intercept in the model always because the RLL value is always larger when the intercept is included. This makes sense with an unbalanced sample. While keeping the intercept in the model may not have substantial effects in a completely balanced sample with larger numbers of selected features, this example in Fig. A7.1(a) and (b) shows that it can cause bias problems and convergence failures with smaller numbers of features as is often the case in Explicit RELR.

Although it was not obvious originally, it does not make sense with completely balanced sample not to include the intercept. Completely balanced samples are highly unlikely to occur in simple random sampling. Yet, in stratified sampling that balances perfectly the two binary groups in logistic regression by randomly selecting an equal number of observations in each binary outcome, balanced samples happen by design. An Explicit RELR model that excludes the intercept and which is based upon a completely balanced stratified sample has especially important application in causal feature learning. This is because the model will not be dependent upon how the variance for the t-value used in the error model is computed or any biasing

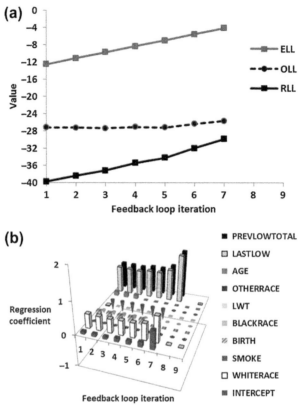

Figure A7.1 Explicit RELR feature selection to predict Low Birth Weight exactly like in Fig. 3.4(a) and (b) except that the intercept is now in the model as a necessary feature. Figure A7.1(a) shows the trajectory of the Explicit RELR computations in terms of the RLL (RELR Log Likelihood), ELL (Error Log Likelihood) and OLL (Observation Log Likelihood) values across nine feedback iterations. Figure A7.1(b) shows the RELR regression coefficients across these same nine feedback iterations. Values for the last two iterations are not shown because the model failed to converge in those last two iterations due to parameter instability.

effects due to the inclusion of an intercept in a balanced sample, which makes the selected features and regression coefficients more interpretable.

The rationale for a zero intercept in completely balanced models is also related to how RELR handles missing values. Given standardized variables, the expected value when all independent variable feature values are missing and all missing status indicator effects are zero or sum to zero is 0. Such a scenario with large numbers of missing feature values could happen in a scoring model for example, where an important feature is simply missing in most or all values due to data collection error. This has the same effect as no imputation or not adding any information to the logistic regression model with missing values provided that the intercept is zero. This is because $\exp(0)/(1 + \exp(0))$ gives a value of 0.5 which is exactly the expected value of the regression when all features have values of zero or at least their regression effects sum to zero and the intercept is also not included in the model built from a perfectly balanced sample. Only in the special case where there is a completely balanced sample with equal strata of target and nontarget outcomes and no intercept will this value of 0.5 make sense. On the other hand, any value other than the prior expected value of 0.5 which will occur with a nonzero intercept would bias the model and return biased parameters and predictions.

In practice, the problem of a nonzero intercept in perfectly balanced samples is only usually noticeable with few selected features in a model. This is because the effect of RELR's error model is to yield an equal probability of positive and negative errors across features, which largely removes this biasing problem when there are more features in a model. This is observed in that the parameters in the model in Fig. A7.1(b) are fairly similar to those in Fig. 3.4(b) until the 7th step which has few remaining features and is where the intercept gets very large.

A8. DETAILED STEPS IN RELR'S CAUSAL MACHINE LEARNING METHOD

This causal hypothesis generation and testing process based upon observation data in RELR ultimately uses a matched sample method to test the hypotheses which is called RELR's outcome score matching method. In the case that a user has previous hypotheses, Step 1 can be modified to involve those hypotheses whether they always include or exclude certain features. Unless those previous hypotheses exist because they were supported in previous well controlled causal tests, that may negate one of the prime advantages of this process which is that the features that are selected as worthy of being tested are not biased by subjective human choices. There is always a chance that hypotheses will be supported by chance. So, at least at the beginning of a research project where clear hypotheses have not yet been formed and causal testing has not yet occurred, a more mechanical, objective and unbiased approach to hypothesis generation might be preferred as argued in this book. This RELR causal machine learning process is summarized in the following 10-step process which is also depicted below graphically in Fig. A8.1:

1. Build an Explicit RELR model (that is a balanced zero intercept model if the outcome is binary) to generate causal hypotheses.

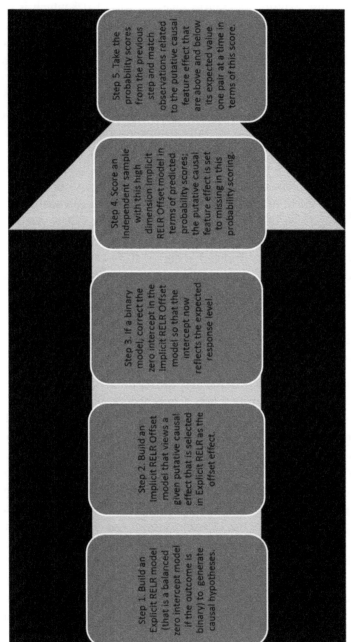

Figure A8.1 RELR's causal machine learning method.

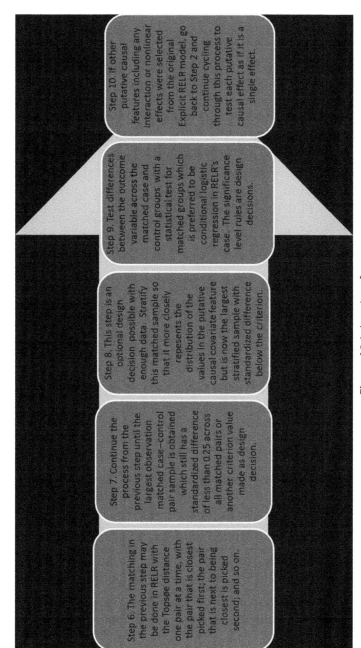

Figure A8.1 *(continued).*

2. Build an Implicit RELR offset model that views a given putative causal effect that is selected from the balanced Explicit RELR zero intercept model as the offset effect, where the offset is the product of the regression coefficient and selected feature value for each data observation. This offset variable will have a constant coefficient of 1 for each data observation and 0 for each pseudo-observation in the offset regression model, so pseudo-observations corresponding to the offset do not influence the RELR error model. This offset regression ideally is a very high-dimension model that would include as covariates all candidate features that resulted from feature reduction before the Explicit RELR model build with high-dimension data. The identical balanced sample used to train the Explicit RELR model should be used to build this Implicit RELR offset model.

3. If necessary, correct the zero or excluded intercept in the Implicit RELR offset model so that the intercept now reflects the expected response level in accordance with the standard method in binary logistic regression.[10]

4. Score an independent sample for predicted probabilities with this high-dimension Implicit RELR offset model. This independent sample necessarily may be at a future time point for applications with potentially time-varying causal effects which will need to assume that causal features precede effects that they cause. The putative causal feature effect is set to missing in this probability scoring.

5. Take the probability scores from the previous step and match observations related to the putative causal feature effect that are above and below its expected one pair at a time in terms of this score which reflects the probability of the outcome given all other covariates when the putative causal effect is set to its expected value which is typically zero with RELR's standardized features, although a user design decision may override this choice.

6. The matching in the previous step may be done in RELR with the Topsøe distance,[11] but any similar measure which gives comparable results in matching two probability distributions in the regions where there are smallest differences may give similar results as long as this design decision is made before viewing the outcome data. The matching should be done one pair at a time with the pair that is closest picked first; the pair that is next to being closest is picked second, and so on as shown in Table 4.1.

7. Continue the process from the previous step until the largest observation matched sample is obtained which still has a standardized difference for the covariate outcome probability scores of less than 0.25 across all matched pairs. Another value of the standardized difference is possible, but this should be made as a design decision before viewing the outcome results.

8. This step is an optional design decision that becomes possible with larger samples of data, but which should be decided in detail before viewing the outcome results. Instead of uniformly sampling the feature values in lockstep with the procedure outlined in Steps 6 and 7, it may be beneficial instead to stratify that sample so that it more closely matches the range of the feature values. So if 2/3 of the case feature values are between 0 and 1 in terms of their standardized values, then a fraction as close as possible to 2/3 of case sample observations could also have their feature values in that range and rest would have their feature values above that range for example. Such a sampling should be done randomly within each stratum so as not to bias that stratum. The largest such stratified sample that still has a standardized difference of covariate outcome probability scores below the criterion threshold should then be chosen as the case–control sample.

9. Test differences between the outcome variable across the matched case and control groups with an appropriate statistical test for matched groups which is preferred to be conditional logistic regression in RELR's case, but may be another test such as McNemar's test of correlated proportions as long as that design decision is made before viewing the outcome.

10. If other putative causal features including any interaction or nonlinear effects were selected from the original Explicit RELR, go back to Step 1 and continue cycling through this process to test each putative causal effect. Note that any given putative causal feature will serve as a covariate that is to be controlled for all other putative causal features. Note that all main effect, interaction and nonlinear effects related to the same variable should be tested separately, as they can have separate meaning as causal features.

One final point concerns the restriction to 1:1 paired sample matching. This is necessary because the matched sample process can become much more complex and possibly arbitrary with *m:n* matching, as order effects might play a role and different distance measures also may have an impact. A 1:1 matched sample is a well-controlled sample that should have very general application, so this restriction should not be seen as severely restricting its application except in the case of extremely small samples where $m{:}n$ matching might give noticeably more reliable estimates. A 1:1 matched sample also mimics common randomized placebo-controlled trial designs where there is also typically a 1:1 match in the exposure and control observations.

Preface

1. Rice, D.M. and Hagstrom, E.C. "Some evidence in support of a relationship between human auditory signal-detection performance and the phase of the alpha cycle", *Perceptual and Motor Skills* 69 (1989): 451–457.
2. Rice, D.M., Buchsbaum, M.S., Starr, A., Auslander, L., Hagman, J. and Evans, W.J. "Abnormal EEG slow activity in left temporal areas in senile dementia of the Alzheimer type", *Journals of Gerontology: Medical Sciences* 45(4) (1990): 145–151. The critical refining aspect was the use of the average recording reference, which gave a purer view of the potential field and the abnormality in the temporal lobes.
3. Rice, D.M., Buchsbaum, M.S., Hardy, D. and Burgwald, L. "Focal left temporal slow EEG activity is related to a verbal recent memory deficit in a nondemented elderly population", *Journals of Gerontology: Psychological Sciences* 46(4) (1991): 144–151.
4. Scoville, W.B. and Milner, B. "Loss of recent memory after bilateral hippocampal lesions", *Journal of Neurology, Neurosurgery and Psychiatry* 20(1) (1957): 11–21.
5. Boon, M.E., Melis, R.J.F., Rikkert, M.G.M.O. and Kessels, R.P.C. "Atrophy in the medial temporal lobe is specifically associated with encoding and storage of verbal information in MCI and Alzheimer patients", *Journal of Neurology Research* 1(1) (2011): 11–15.
6. Henneman, W.J.P., Sluimer, J.D., Barnes, J., van der Flier, W.M., Sluimer, I.C, Fox, N.C., Scheltens, P., Vrenken, H. and Barkhof, F. "Hippocampal atrophy rates in Alzheimer disease: added value over whole brain volume measures", *Neurology* 72 (2009): 999–1007.
7. Duara, R., Loewenstein, D.A., Potter, E., Appel, J., Greig, M.T., Urs, R., Shen, Q., Raj, A., Small, B., Barker, W., Schofield, E., Wu, Y. and Potter, H. "Medial temporal lobe atrophy on MRI scans and the diagnosis of Alzheimer disease", *Neurology* 71(24) (2008): 1986–1992.
8. McKhann, G.M., Knopman, D.S., Chertkow, H., Hyman, B.T., Jack, C.R. Jr., Kawas, C.H., Klunk, W.E., Koroshetz, W.J., Manly, J.J., Mayeux, R., Mohs, R.C., Morris, J.C., Rossor, M.N., Scheltens, P., Carrillo, M.C., Thies, B., Weintraub, S. and Phelps, C.H. "The diagnosis of dementia due to Alzheimer's disease: recommendations from the National Institute on Aging-Alzheimer's Association workgroups on diagnostic guidelines for Alzheimer's Disease", *Alzheimer's and Dementia* 7(3) (2011): 263–269.
9. Kuhn, T.S. *The Structure of Scientific Revolutions* (1st ed. 1962, 4th ed. 2012: Chicago, University of Chicago Press).
10. Marchione, M. "Alzheimer drug shows some promise in mild disease", Bloomberg Business Week, October 8, 2012.
11. Khosla, D., Singh, M. and Rice, D.M. "Three dimensional EEG source imaging via the maximum entropy method", *IEEE Nuclear Science Symposium and Medical Imaging Record* 3 (1995): 1515–1519.

Chapter 1

1. Leibniz, G. 1685. *The Art of Discovery*; In Wiener, Philip, *Leibniz: Selections*, (1951: Scribner).
2. Bardi, J.S. *The Calculus Wars: Newton, Leibniz and the Greatest Mathematical Clash of all Time* (2006: New York, Avalon Group/Thunder's Mouth Press).

3. Ibid.

4. Ibid, the word "calculus" is derived from the Greek word for a small stone that was used to make calculations. As Bardi points out, this was Leibniz's own invented term which Newton did not use. The word "ratio" has origins in the Latin word for reason. According to Merriam-Webster, the noun "ratiocination" means the process of exact thinking or a reasoned train of thought. http://www.merriam-webster.com/dictionary/ratiocination, so a *Calculus Ratiocinator* might be taken to be a device that produces exact thinking through computation. Ratiocinator is pronounced by combining the words "ratio", "sin", "ate" and "or".

5. Leibniz, G. 1666. *On the Art of Combination*; In Parkinson, G.H.R., Logical Papers. (1966: Oxford: Clarendon Press).

6. Leibniz, *The Art of Discovery*, op. cit.

7. Today's widely used predictive analytics methods include Support Vector Machines, Artificial Neural Networks, Decision Trees, Standard Logistic and Ordinary Least Squares Regression, Genetic Programming, Random Forests, Ridge, LASSO, and LARS Regression and many others as reviewed through the course of this book.

8. Public domain image: http://en.wikipedia.org/wiki/File:Gottfried_Wilhelm_von_Leibniz.jpg.

9. Dejong, P. "Most data isn't big and businesses are wasting money pretending it is", *Quartz*, May 6, 2013, http://qz.com/81661/most-data-isnt-big-and-businesses-are-wasting-money-pretending-it-is/.

10. Wolpert, D.H. "The Lack of Apriori Distinctions between Learning Algorithms", *Neural Computation* 8(7) (1996): 1341–1390.

11. Wolpert, D.H. and Macready, W.G. "No Free Lunch Theorems for Optimization", *IEEE Transactions on Evolutionary Computation* 1(1) (1997): 67–82.

12. Nelder, J.A. "Functional Marginality and Response Surface Fitting", *Journal of Applied Statistics* 27 (2000): 109–112. This scaling problem was passionately described by Nelder throughout his career, and this referenced article is just one example.

13. Standardized variables with a mean of 0 and a standard deviation of 1 are orthogonal when they are uncorrelated.

14. Breiman, L. "The Little Bootstrap and Other Methods for Dimensionality Detection in Regression: X-Fixed Prediction Error", *Journal of the American Statistical Association* 87 (1992): 738–754.

15. Austin, P.C. and Tu, J.V. "Automated Variable Selection Methods for Logistic Regression Produced Unstable Models for Predicting Acute Myocardial Infarction Mortality", *Journal of Clinical Epidemiology* 57 (2004): 1138–1146.

16. The Akaike Information Criterion and Bayesian Information Criterion are two such parsimony criteria commonly used in logistic regression.

17. Wickham, H. "Exploratory Model Analysis", *JSM Proceedings* 2007; available online at http://had.co.nz/model-vis/2007-jsm.pdf.

18. Austin and Tu, op. cit. There was only a very slight improvement in the reliability of variable selections when experts also helped to select variables.

19. Shi, L. et al. "The MicroArray Quality Control (MAQC)-II Study of Common Practices for the Development and Validation of Microarray-Based Predictive Models", *Nature Biotechnology* 28(8) (2010): 827–838.

20. Still, J. et al. "Feature-Weighted Linear Stacking", 2009; available online at www.arxiv.org.

21. Some of these problems are due to a tendency to give biased estimates when data are from different types such as categorical and continuous variables. See Strobl, C. et al. "Bias in Random Forest Variable Importance Measures: Illustrations, Sources and a Solution", *BMC Bioinformatics* 8 (2007): 25. Important variables are also more likely to be correlated predictors. The proposed solution to modify the original Random Forest method helps somewhat, but it does not eliminate this problem with multicollinear

variables. See Strobl, C. et al. "Conditional Variable Importance for Random Forests", *BMC Bioinformatics* 9 (2008): 307.

22. Seni, G. and Elder, J. *Ensemble Methods in Data Mining: Improving Accuracy through Combining Predictions*, (2010: Morgan and Claypool Publishers).

23. Miller, G.A. "The Magical Number Seven, Plus or Minus Two: Some Limits on Our Capacity for Processing Information", *Psychological Review* 63 (1956): 81–97. It might be more accurate to say that there are these small number limitations in the conscious objects, as a given object might be made from a larger number of features. Still the brain only appears to be able to hold a limited set of features in consciousness.

24. Dew, I.T. and Cabeza, R. "The Porous Boundaries between Explicit and Implicit Memory: Behavioral and Neural Evidence", *Annals of the New York Academy of Sciences* 1224 (2011): 174–190.

25. Baddeley, A.D. and Hitch, G.J. "Working Memory", In Bower, G.A. (Ed.), *The Psychology of Learning and Motivation: Advances in Research and Theory*, Vol. 8, pp. 47–89, (1974: New York, Academic Press). Baddeley, A.D. "The Episodic Buffer: A New Component of Working Memory?" *Trends in Cognitive Sciences* 4(11) (2000): 417–423.

26. Fell, J. and Axmacher, N. "The Role of Phase Synchronization in Memory Processes", *Nature Reviews Neuroscience* 12 (2011):105–118.

27. Craik, F.I.M. and Tulving, E. "Depth of Processing and the Retention of Words in Episodic Memory", *Journal of Experimental Psychology: General* 104(3) (1975): 268–294.

28. Warrington, E.K. and Weiskrantz, L. "Amnesic Syndrome: Consolidation or Retrieval?" *Nature* 228 (1970): 628–630.

29. Kahneman, D. *Thinking, Fast and Slow* (2011: New York, Farrar, Straus and Giroux).

30. Duhigg, C. *The Power of Habit* (2012: New York, Random House).

31. Stefanacci, L., Buffalo, E.A., Schmolck, H. and Squire, L.R. "Profound Amnesia after Damage to the Medial Temporal Lobe: A Neuroanatomical and Neuropsychological Profile of Patient E.P.", *Journal of Neuroscience* 20(18) (2000): 7024–36.

32. Graybiel, A.M. "The Basal Ganglia and Chunking of Action Repertoires", *Neurobiology of Learning and Memory* 70 (1998): 119–36.

33. In the basal ganglia, this predictive model learning occurs in parallel at each projection stage across a large population of basal ganglia neurons, and this circuitry also includes local feed-forward interneuron connections. See Graybiel, A.M. "Network-Level Neuroplasticity in Cortico-Basal Ganglia Pathways", *Parkinsonism and Related Disorders* 10 (2004): 293–296.

34. Dew and Cabeza, *Annals of the New York Academy of Sciences*, op. cit.

35. Although unlike more explicit memories, familiarity memory does not appear to be disrupted by selective hippocampus injury even though select human hippocampus neurons actually show rapid old–new familiarity discriminations but these are probably too fast for conscious recollection. See Rutishauser, U., Mamelak, A.N. and Schuman, E.M. "Single-Trial Learning of Novel Stimuli by Individual Neurons of the Human Hippocampus–Amygdala Complex", *Neuron* 49(6) (2006): 805–813.

36. Many cognitive neuroscientists believe that familiarity shares an underlying mechanism with repetition priming because studies of the temporal course of the brain's electric activity during recognition memory tasks suggest that they both operate in the same rapid response time frame and are difficult to dissociate, but an alternative view is that these recognition memory processes simply operate concurrently. See Dew and Cabeza, *Annals of the New York Academy of Sciences*, op. cit.

37. An interesting side note is that even though stacked ensemble models built from hundreds of submodels were most accurate in the Netflix competition, a linear ensemble blend of just two submodels, one built from the Singular Value Decomposition (SVD) method and the other built from Restricted Boltzmann Machines method, was accurate enough and far simpler to implement. So, this far simpler linear

ensemble blend of just two submodels was put into production in the recommendation system at Netflix. See Amatriain, X. and Basilico, J. "Netflix Recommendations: Beyond the 5 Stars (Part 1)", *Netflix Tech Blog*, April 6, 2012.

38. Breiman, L. "Statistical Modeling: The Two Cultures", *Statistical Science* 16(3) (2001): 199–231.

39. Shmueli, G. "To Explain or To Predict?" *Statistical Science* 25(3) (2010): 289–310.

40. Derman, E. *Models Behaving Badly: Why Confusing Illusion with Reality Can Lead to Disaster, on Wall Street and in Life* (2011: New York, Free Press).

41. Buzsáki, G. *Rhythms of the Brain* (2006: Oxford, Oxford University Press).

42. Buzsáki, *Rhythms of the Brain*, op. cit. See p. 297 discussion of how only small fraction of cells in hippocampus fire in any given place in an environment and pp. 284–287 for discussion of feedback in medial temporal lobe and hippocampus, along with entorhinal reciprocal connection to neocortex.

43. Karlsson, M.P. and Frank, L.M. "Network Dynamics Underlying the Formation of Sparse, Informative Representations in the Hippocampus", *Journal of Neuroscience* 28(52) (2008): 14271–14281.

44. There are important exceptions to when humans with normal cognitive ability can all agree on shared episodic experiences. These exceptions involve more remote experiences when suggestion is also involved to create false memories. Childhood memories are especially susceptible to these false suggestion effects. See Loftus, E. "Creating False Memories", *Scientific American* 277 (1997): 70–75.

45. Buckner, R.L. and Carroll, D.C. "Self-Projection and the Brain", *Trends in Cognitive Sciences* 11(2) (2006): 49–57.

46. Lehn, H., Steffenach, H.A., van Strien, N.M., Veltman, D.J., Witter, M.P. and Håberg, A.K. "A Specific Role of the Human Hippocampus in Recall of Temporal Sequences", *The Journal of Neuroscience* 29(11) (2009): 3475–3484.

47. It has also been pointed out that it is misleading to characterize medial temporal lobe patients as having conscious recollection problems because they can clearly recollect remote experiences. They also show problems with tasks where stimuli relations are learned unconsciously, so their memory problems may be better characterized as problems in remembering relations between stimuli. In any case, the hippocampus does seem necessary to experience consciously recollected relations between recent events. See Eichenbaum, H. and Cohen, N.J. *From Conditioning to Conscious Recollection: Memory Systems of the Brain* (2004: Oxford, Oxford University Press).

48. The 2011 Data Mining Survey by Rexer Analytics concluded that while a wide variety of methods are used in predictive analytics, regression and decision tree methods are used most frequently. See http://www.rexeranalytics.com.

49. See discussion of this rise in popularity in the preface in Hosmer, D.W. and Lemeshow, S. *Applied Logistic Regression* (2000: New York, Wiley, 2nd Edition). They mention how there was an "explosion" in the usage of logistic regression over the period from the later 1980s when their first edition came out through 2000 when their second edition appeared. Every indication is that this rapid rise in popularity has continued.

50. Peng, C.Y.J. and Po, T.S.H. "Logistic Regression Analysis and Reporting: A Primer", *Understanding Statistics* (2002) 1: 31–70.

51. See Aldrich, J.H. *Linear Probability, Logit, and Probit Models*, (Sage Publications, London, 1984).

52. Ibid.

53. Hosmer, D.W. and Lemeshow, S. *Applied Logistic Regression*, (2000: New York, Wiley).

54. Allison, P.D. "Discrete-Time Methods for the Analysis of Event Histories", *Sociological Methodology*, Vol. 13, (1982), pp. 61–98. Chapter 3 will detail some of the nuances of using logistic regression in survival analysis, and why it has not been widely used in this case.

55. Mount, J. "The Equivalence of Logistic Regression and Maximum Entropy Models", online at http://www.win-vector.com/dfiles/LogisticRegressionMaxEnt.pdf.

56. There are many reasons reviewed through this book that the error-in-variables modeling in logistic regression is a much simpler problem than the linear regression case, even though linear regression has received far more study. For an overview of this method applied to linear regression see Gillard, J. "An Overview of Linear Structural Models in Errors in Variables Regression", *Revstat – Statistical Journal* 8(1) (2010): 57–80.

57. Hastie, T., Tibshirani, R. and Friedman, J. *Elements of Statistical Learning: Data Mining, Inference, and Prediction*, (2011: Springer-Science, New York) 5th Edition, online publication.

58. For a concise summary of how modelers typically attempt to overcome some of the problems that result in the LASSO, see blogger Karl R.'s comment following Andrew Gelman's blog: "Tibshirani Announces New Research Result: A Significance Test for the LASSO", on March 18, 2013, 10:55 am. The most useful method is simply to decorrelate independent variables so that they are no longer multicollinear—standard approaches are to drop variables or average variables through methods like Principal Components Analysis. However, because one has to remove multicollinearity to make optimal use of LASSO, this suggests that LASSO is not a very viable solution to the multicollinearity error problems. http://andrewgelman.com/2013/03/18/tibshirani-announces-new-research-result-a-significance-test-for-the-lasso/. The Elastic Net method which combines LASSO and Ridge Regression with the added cost of returning many more nonzero variables than LASSO might be a more viable way to handle multicollinearity error problems, although a study reviewed later in the book—Haury, A.-C., Gestraud, P., Vert, J.-P. (2011). "The Influence of Feature Selection Methods on Accuracy, Stability and Interpretability of Molecular Signatures", *PLoS One* 6(12): e28210. http://dx.doi.org/10.1371/journal.pone.0028210—suggests that methods closer to RELR are more effective.

59. This is based upon the maximum entropy subject to linear constraints method first used by Boltzmann in the original ensemble modeling in the nineteenth century, as the maximum entropy subject to constraints and maximum likelihood solutions are identical in logistic regression. The original usage of maximum entropy subject to constraints methods might have been by Boltzmann. Note that if one assumes that one must perform some sort of averaging of elementary models to qualify for the definition of an ensemble method, then technically Implicit RELR is not an ensemble method because it returns a solution across many redundant features automatically. However, this Implicit RELR solution is actually the most probable weighted average that would be obtained by averaging many simpler sparse solutions, so it returns a solution that would have been formed through optimal weighted averaging of elementary models. RELR's Implicit "ensemble" method is actually similar to the usage of the term ensemble in statistical mechanics. This original usage of the concept of the ensemble is actually based upon maximum entropy statistical mechanics modeling where there are a set of microstates corresponding to possible molecular configurations, and the maximum entropy solution describes the most likely ensemble of microstates in terms of macro feature constraints such as involving temperature, pressure and energy.

60. In the commercial software implementation of RELR as a SAS language macro, Explicit RELR is currently called Parsed RELR and Implicit RELR is called Fullbest RELR. The generic explicit and implicit terms are used throughout the book because they are more descriptive and more general.

61. Mach, E. "The Guiding Principles of My Scientific Theory of Knowledge and Its Reception by My Contemporaries", In *Physical Reality*, Edited by Toulmin, S. (1970: New York, Harper Torchbooks).

62. Skinner, B.F. About Behaviorism (1970: New York: Vintage).
63. Mach, E. *The Science of Mechanics: A Critical and Historical Account of its Development* (1919: Chicago, Open Court Publishing), originally published in 1883.
64. Kelly, J. *Gunpowder: Alchemy, Bombards, and Pyrotechnics: The History of the Explosive that Changed the World* (2005: New York, Perseus Books Group).

Chapter 2

1. Boltzmann, L. from *Populäre Schriften*, Essay 3. Address to a Formal meeting of the Imperial Academy of Science, May 29, 1886. In Brian McGuinness (ed.), *Ludwig Boltzmann: Theoretical Physics and Philosophical Problems, Selected Writings* (1974: New York, Springer).
2. Perrin, J.B. "Discontinuous Structure of Matter", Lecture given on December 11, 1926, *Nobel Lectures, Physics 1922–1941*, (1965: Amsterdam, Elsevier).
3. As reviewed by Frigg and Werndl, it was eventually pointed out that Boltzmann's concept that systems would naturally evolve toward higher entropy would only happen with further ergodic assumptions, and even in this case the evolution would only then stay close to its maximum entropy value most of the time. In thermodynamics, ergodicity can be roughly taken to mean that the system has the same behavior when averaged over time as when averaged over space, so that particles will have an equal probability of being in all possible position and momentum states over long periods of time. Frigg, R. and Werndl, C. "Entropy—A Guide for the Perplexed". In *Probabilities in Physics*, Edited by Beisbart, C. and Hartmann, S. (2010: Oxford, Oxford University Press).
4. Shannon, C. and Weaver, W. *The Mathematical Theory of Communication*, (1949: Urbana, University of Illinois Press). The Shannon expression is opposite in sign from Boltzmann's entropy because probabilities are used in this Shannon expression which reverse the sign of the logarithm. When all probabilities $p(j)$ are equal to a constant, then Shannon's expression reduces to $-\log W$ where W is the product of the individual probabilities $p(j)$ across all j, which is Boltmann's entropy with the scaling constant $k = -1$. Shannon's expression is more general than Boltzmann's because it does not require an arbitrary scaling constant specific to a given statistical mechanics implementation. The base e logarithm form is used in this book because it gives equivalent solutions as maximum likelihood logistic regression.
5. Jaynes, E. "Information theory and statistical mechanics", *Physical Review* 106 (1957): 620–630.
6. This Jaynes Maximum Entropy Principle has not been without controversy. See Howson, C. and Urbach, P. Scientific Reasoning: The Bayesian Approach, (2006: La Salle, Open Court). They critique the Jaynes notion that objective prior probability can be obtained by maximizing entropy only on the basis of historical data and known constraints. Their critique centers on problems with continuous information entropy definitions, which may not entirely apply to the discrete Shannon entropy that is the basis of the logistic regression in this book. These authors are subjective Bayesians as opposed to Jaynes' objective Bayesian approach, and they view probability as a measure of degree of belief. This distinction will be discussed in detail in the next chapter.
7. Lucas, K. and Roosen, P. "Transdisciplinary Fundamentals" In *Emergence, Analysis and Evolution of Structures: Concepts and Strategies Across Disciplines*. Edited by Lucas, K. and Roosen, P. (2010, Berlin-Heidelberg, Springer-Verlag): 5–73.
8. Public domain image: http://en.wikipedia.org/wiki/File:Bentley_Snowflake9.jpg.
9. The Hodgkin and Huxley model uses an equation that describes how the concentration gradient in the distribution of a charged molecule across a divide such as a cell membrane is related to an electrostatic potential gradient. This effect can be understood

as a mechanism that maximizes entropy subject to constraints in accordance with the second law of thermodynamics. This idea is basic to all concepts of how graded potentials arise across cell membranes, along with what causes ions to move across synaptic gates that temporarily open to produce ionic currents and corresponding graded potential changes that ultimately influence the probability of spiking at the soma as reviewed in the next chapter. See Hodgkin, A.L. and Huxley, A.F. "A Quantitative Description of Membrane Current and its Application to Conduction and Excitation in Nerve", *Journal of Physiology* 117(1952): 500–544.

10. Golan, A., Judge, G. and Perloff, J.M. "A Maximum Entropy Approach to Recovering Information from Multinomial Response Data", *Journal of the American Statistical Association* 91 (1996): 841–853.
11. Allison, P.D. "Convergence Failures in Logistic Regression", *SAS Global Forum*, Paper 360, (2008): 1–11.
12. Hosmer, D.W. and Lemeshow, S. *Applied Logistic Regression*, op. cit.
13. Aldrich, J.H. and Nelson, F.D. *Linear Probability, Logit, and Probit Models*, (1984, London, Sage Publications).
14. Mount, J. "The Equivalence of Logistic Regression and Maximum Entropy Models", op. cit.
15. Egan, A. "Some Counterexamples to Causal Decision Theory", *The Philosophical Review* 116(1) (2007): 93–114.
16. McFadden, D. (1974). "Conditional Logit Analysis of Qualitative Choice Behavior", In Zarembka, P. (ed.) *Frontiers in Econometrics* (1974: New York, Academic Press) 105–142.
17. Train, K. *Discrete Choice Methods with Simulation*, (2nd Edition 2009: Cambridge, Cambridge University Press).
18. Ibid.
19. Ibid.
20. Hosmer, D.W. and Lemeshow, S. *Applied Logistic Regression*, op. cit.
21. Wall, M.M., Dai, Y. and Eberly, L.E. "GEE Estimation of a Misspecified Time-Varying Covariate: An Example with the Effect of Alcoholism on Medical Utilization", *Statistics in Medicine* 24(6) (2005): 925–939.
22. These are not really random effects but instead random intercepts, but the usage of the term is still widely applied to this meaning. On the other hand, the term "random coefficient" is often applied when coefficients in regression effects are also modeled to vary randomly across the clustering units.
23. One recently proposed alternative option that is implemented in linear regression is that the controversial random effects assumption that the random error effects across the clustering units are uncorrelated with the nested lower level explanatory variable effects is forced through design. See Bartels, B.L., "Beyond 'Fixed versus Random Effects': A Framework for Improving Substantive and Statistical Analysis of Panel, Time-Series Cross-Sectional, and Multilevel Data", online publication originally presented at the Faculty Poster Session, *Political Methodology Conference*, Ann Arbor, MI, July 9–12, 2008. This will avoid explanatory regression coefficients that are biased by such correlation, but it still may not be appropriate depending upon whether the random effects assumption is reasonable to begin with. In some cases, it may be reasonable to believe that the unobserved factors in the clustering units may be expected to correlate with the explanatory variables, so the random effects assumption is simply not appropriate in those cases.
24. Train, K. Discrete Choice Methods with Simulation, op. cit.
25. Henscher, D.A. and Greene, W.H. "The Mixed Logit Model: The State of Practice and Warnings for the Unwary", online paper.
26. Ibid.
27. Ibid.

28. Hosmer, D.W. and Lemeshow, S. *Applied Logistic Regression*, op. cit.

29. Gillard, J. "An Overview of Linear Structural Models in Errors in Variables Regression", op. cit.

30. For example, see review by Rhonda Robertson Clark, "The Error-in-Variables Problem in the Logistic Regression Model", Doctoral Dissertation, University of North Carolina-Chapel Hill (1982).

31. Ibid.

32. However, significant coverage anomalies still exist in small samples in standard measures of such binomial proportion error. See Agresti, A. and Coull, B.A. "Approximate is Better than 'Exact' for Interval Estimation of Binomial Proportions", *American Statistician* 52 (1998): 119–126.

33. Gillard, J. "An Overview of Linear Structural Models in Errors in Variables Regression", op. cit.

34. RELR is only applied to the cases of binary and ordinal regression, but any model with multiple qualitative categories that would be modeled with multinomial standard logistic regression also can be modeled as separate binary RELR models. The main body of the book presents the RELR binary model in conjunction with the previous sections, but Appendix A1 describes the special case of RELR ordinal regression.

35. For another treatment of error modeling in maximum entropy and logistic regression see Golan, A., Judge, G. and Perloff, J.M. "A Maximum Entropy Approach to Recovering Information from Multinomial Response Data", op. cit. Golan, A., Judge, G. and Perloff, J.M. *Maximum Entropy Econometrics*, (1996: New York, Wiley). The specification of the RELR error differs from this previous work in that RELR assumes that the log odds of error is approximately inversely proportional to t, where t can be defined for each independent variable feature as discussed in Appendix A1.

36. Luce, R.D. and Suppes, P. "Preference, Utility and Subjective Probability", In Luce, R.D., Bush, R.R. and Galanter, E. (eds), *Handbook of Mathematical Psychology* 3 (1965: New York, Wiley and Sons) 249–410.

37. McFadden, D. "Conditional Logit Analysis of Qualitative Choice Behavior", op. cit.

38. Davidson, R.R. "Some Properties of Families of Generalized Logistic Distributions", *Statistical Climatology, Developments in Atmospheric Science*, 13, Ikedia, S. et al. (ed.), (1980: New York, Elsevier).

39. George, E.O. and Ojo, M.O. "On a Generalization of the Logistic Distribution", *Annals of the Institute of Statistical Mathematics* 32(2, A) (1980): 161–169. George and Ojo showed that when the standardized logit coefficient is multiplied by a scaling constant, these two probability distributions differ by less than 0.005% with a small number of degrees of freedom of less than 10 and this difference rapidly goes to zero with larger numbers of degrees of freedom.

40. This was originally shown in Rice, D.M. "Generalized Reduced Error Logistic Regression Machine", *Section on Statistical Computing—JSM Proceedings* (2008): 3855–3862.

41. When only a prediction is desired, which is a typical application of Implicit RELR, balancing samples may not add much because models are not interpreted in terms of explanations. Still there can be an advantage even here in that fewer total observations are used in balanced samples. Yet, whenever Explicit RELR is computed to generate an explanation, balanced samples are strongly recommended because with no missing data the solution will not be dependent upon how the variance estimates in t values were computed. The formula for t values that is based upon Pearson correlations shown in the Appendix is used in the error model expression and feature reduction unless special conditions apply. A Student's t value that assumes equal variance in two independent groups also gives approximately the same error modeling estimates here as does Welch's t in balanced binary RELR samples without missing data. Yet, Welch's t is always used in some special conditions as noted in the Appendix, as for example this can give

noticeably more stable Explicit RELR feature selections when missing feature values are imbalanced across binary outcomes.

42. This issue of the Nelder marginality principle was discussed in the first chapter, Nelder, J.A. "Functional Marginality and Response Surface Fitting", op. cit.

43. Moosbrugger, H., Schermelleh-Engel, K., Kelava, A., Klein, A.C. "Testing Multiple Nonlinear Effects in Structural Equation Modeling: A Comparison of Alternative Estimation Approaches". Invited Chapter in Teo, T. & Khine, M.S. (eds), *Structural Equation Modelling in Educational Research: Concepts and Applications* (In press, Rotterdam, Sense Publishers).

44. Rice, D.M. "Generalized Reduced Error Logistic Regression Machine", op. cit.

Chapter 3

1. Fienberg, S. "When Did Bayesian Inference Become 'Bayesian'?" *Bayesian Analysis* 1 (1) (2003): 1–41.

2. Frigg, R. and Werndl, C. "Entropy—A Guide for the Perplexed", op. cit.

3. Ibid.

4. Tversky, A. and Kahneman, D. "Judgment under Uncertainty: Heuristics and Biases", *Science* 185 (1974): 1124–1131.

5. Ibid.

6. The KL divergence is not a metric because it does not exhibit the triangular property which determines that the shortest path between two points is a straight line.

7. Eguchi, S. and Copas, J. "Interpreting Kullback–Leibler Divergence with the Neyman–Pearson Lemma", *Journal of Multivariate Analysis* 97 (2006): 2034–2040.

8. Eguchi, S. and Kano, Y. "Robustifing Maximum Likelihood Estimation" (2001): online publication.

9. Allison, P.D. *Sociological Methodology*, op. cit.

10. Hedeker, D. "Multilevel Models for Ordinal and Nominal Variables", *Handbook of Multilevel Analysis*, edited by Jan de Leeuw and Erik Meijer, 2007 Springer, New York pp. 239–276. Hedeker brings multilevel modeling into his survival analysis methods, but as will be reviewed in Chapter 4, RELR does not require multilevel parameters so the parts of his method that relate to multilevel parameters do not apply to RELR.

11. Allison, P.D. *Sociological Methodology*, op. cit. Hedeker, D., op. cit.

12. Allison, P.D. *Sociological Methodology*, op. cit. Hedeker, D., op. cit.

13. Allison, P.D. *Sociological Methodology*, op. cit.

14. Allison, P.D. *Sociological Methodology*, op. cit. Hedeker, D., op. cit.

15. Hedeker, D., op. cit.

16. Hedeker, D., op. cit.

17. RELR will however exhibit the classic almost complete separation problems in high-dimension data relative to the sample size unless the sample is perfectly balanced with no intercept/zero intercept as reviewed in an upcoming example. See Allison, P.D. "Convergence Failures in Logistic Regression", op. cit. for a description of this problem.

18. Maximum entropy and maximum log likelihood return the identical solution which is the maximum probability solution when known constraints are entered into a model. Yet, the meaning of these two measures is different when the constraints are not known. This difference is because these two measures are the same magnitude value but opposite in sign for a given set of feature constraints. So, in both standard logistic regression and RELR, a search across all possible selected feature sets will find that the maximum entropy solution across all possible selections will be a different solution than the maximum log likelihood solution across all possible selected feature sets. In this search across all possible feature sets in standard logistic regression, the maximum entropy solution is a trivial solution with no constraints at all. Thus, nothing is effectively learned by maximizing uncertainty in feature selection there. On the other hand, in the search for maximum log likelihood across all

possible feature sets, the standard logistic regression solution contains all the features no matter how many. Such a solution is usually associated with extreme overfitting and very poor prediction generalization to new data samples unless there are very few total feature constraints in relation to the sample size. The problem is that each additional feature added in standard logistic regression always increases the log likelihood value in the fit to the training sample. So in standard logistic regression, the log likelihood value is not a good measure either that can be used to choose an optimal feature selection set that would generalize well in prediction to new data samples.

19. Seni and Elder, op. cit.
20. While this is implemented as a backward selection process in current software implementations, it is possible that an equivalent process could be implemented as a forward selection process which simply adds features based upon t-value magnitudes while controlling for bias in even versus odd polynomial features. Since the only goal is prediction in this feature selection, using a backward versus forward selection process can give identical selections and prediction. But when they differ in real-world data, it is because backward selection maintains linear features longer in the process which are more prevalent than nonlinear features because binary and category variable cannot be nonlinear. In such cases, it is usually an advantage that linear features are maintained longer because nonlinear features are often likely to be spurious or even falsely positive correlation effects as reviewed in the last chapter.
21. Should a tie between two or more solutions ever occur this method simply selects the solution with the fewest features.
22. Seni, G. and Elder, J., op. cit.
23. Seni, G. and Elder, J., op. cit.
24. One obvious very minor change that may sometimes lead to a slight improvement in observation log likelihood fit would be to allow either an odd- or even-powered polynomial feature to be dropped at each step, but force that both need to occur in every two steps. This may sometimes yield slightly larger observation log likelihood values across all steps than dropping both one odd- and one even-powered polynomial feature at each step, but it requires double the processing. The reason that it has not been implemented in Implicit RELR is because the improvement in the log likelihood is very small and not significant whereas the savings of half the processing effort with the current implementation is substantial.
25. van der Laan, M.J. and Rose, S. "Targeted Learning: Causal Learning for Observational and Experimental Data", (Springer, New York, 2012). See Chapter 3 on Super Learners.
26. Rice, D.M. *JSM Proceedings*, op. cit. Note that the very poor performance of Partial Least Squares (PLS) is more about the fact that many of predictive features were binary features, which is actually a violation of best practice assumptions in PLS and Principal Component Analysis (PCA). Yet, this is a clear example of the problems that arise when restrictive assumptions of standard predictive analytic methods like PCA and PLS are not met.
27. Mitchell, T. "Generative and Discriminative Classifiers: Naïve Bayes and Logistic Regression", Online draft.
28. Haury, A.C., Gestraud, P. and Vert, J.P. "The Influence of Feature Selection Methods on Accuracy, Stability and Interpretability of Molecular Signatures", *PLoS ONE* 6(12) (2011): e28210. http://dx.doi.org/10.1371/journal.pone.0028210.
29. Hand, D.J. "Measuring Classifier Performance: A Coherent Alternative to the Area Under the ROC Curve". *Machine Learning* 77 (2009): 103–123.
30. Hosmer, D.W. and Lemeshow, S., op. cit. They give a more detailed discussion of Wald and Likelihood Ratio test theory and computations in logistic regression along with other similar tests such as the Score test. Yet a very succinct description is that Likelihood Ratios are used to compute such a χ^2 which estimates the effect of removing one

parameter. So, the reduced model needs to be completely nested within the fuller model. That is, everything is the same in both models with the exception that the parameter that is tested is zeroed in the reduced model. In this test, minus two times the difference in log likelihood of the reduced model to the fuller model or $-2LL$ is distributed as a χ^2 variable with one degree of freedom.

31. For example, it is expected that t-values that measure reliability of correlation effects and that are closer to zero would vary less in magnitude across independent samples compared with such larger magnitude t-values. For example, in large enough samples where t-values that reflect the reliability of Pearson correlations would be expected to be stable such as 1000 observations, it would not be surprising if a t-value varied between 248 and 265 across two independent samples because this would translate into a correlation varying between 0.992 and 0.993 which we know intuitively is likely to be rounding error. But it would be surprising if a t-value varied between 0 and 17 across two similar samples, because this would reflect the very unlikely occurrence of a correlation varying between 0 and 0.475 across these two samples.

Fisher developed a test called the z' test to correct for the tendency for Pearson correlations to have differential variability across their range and thus allow correlations to be compared with one another. But here we are talking about how t-values that measure correlation effects would have a different type of variability across their range. Yet, Fisher's test would predict that a correlation that changes between 0 and 0.47 across two independent samples of 1000 observations is a very unlikely occurrence, whereas a fluctuation between 0.992 and 0.993 is a likely occurrence. See Fisher, R.A. "On the 'Probable Error' of a Coefficient of Correlation Deduced from a Small Sample", *Metron* 1 (1921): 3–32.

32. Different types of data in terms of different domains may require somewhat different considerations in reasonable feature reduction short lists for Explicit RELR, but there will be much less variability within a domain. Like Implicit RELR, Explicit RELR's feature selection is clearly multivariate in the sense of being able to handle interaction effects. Yet, unlike Implicit RELR, its regression coefficients in its final models are often not proportional to t-value magnitudes. Still, it does have similarities to Implicit RELR in that the relative magnitude and signs of the regression coefficients are not substantially affected by the presence or absence of other features in the model.

Depending upon the starting point in terms of the magnitude of the t-values that are considered as candidate features, Explicit RELR may miss causal features that have very small univariate effects. This would happen for example with causal features that are almost perfectly negatively correlated to one another if they cancel and have little effect on the dependent variable. But, such extreme cancellation is an unlikely possibility, although it is certainly possible. As shown in the simulations of the last chapter, even when linear and cubic features are simulated to have regression coefficients of the same magnitude but oppositely correlated, the t-values observed in RELR are still substantial with such opposing features and consistent with a Taylor series approximation to the sine function which simply gives less weight to the cubic effect. One may view Explicit RELR as obtaining most likely feature selection under the condition that parsimony is a strong objective, but this does not mean that Explicit RELR will necessarily select causal features.

33. This can be observed in as the regression coefficients shown in Fig. 3.2(a) are from an RELR model with only 176 features whereas those in Fig. 3.2(c) and (d) are from an

RELR model based upon the same data set with 9 features, and standard error *se* values are higher with 9 features so stability *s* or $1/se$ is lower with 9 remaining features.

34. The χ^2 values that are required to compute the magnitude of the derivative in Eqn (3.5b) can be computed using methods such as the Likelihood Ratio, the Score or the Wald approximation. Yet, commercial software implementations of RELR use the Wald approximation to calculate χ^2 and get the corresponding χ value. Note that a very easy way to compute this derivative that avoids the chain rule in the appendix is to realize that $d(\beta s)^2/ds$ is simply $2\beta^2 s$ or $2\beta\chi$, since $s = 1/se$ and $\chi = \beta/se$.

35. Available at http://www.umass.edu/statdata/statdata/data/.

36. In RELR, interaction and nonlinear effects are always computed between standardized variables, and then the interaction or nonlinear feature is itself standardized. This removes the marginality problems related to differential scaling when interaction and nonlinear effects are introduced in regression modeling, as all such effects are on the same relative scale as main effects.

37. King, G. and Zeng, L. "Logistic Regression in Rare Events Data", *Political Analysis* 9 (2001): 137–163. Copy at http://j.mp/lBZoIi. In accord with their Eqn (7) on p. 144 in this article, the adjusted intercept α can be defined as $\alpha = -\ln((1 - \tau)/\tau)$ in the case of a perfectly balanced stratified sample by design, where τ is an estimate of the fraction of 1's that would be found in the population which is gotten from the proportion of 1's prior to sample balancing in the original Explicit RELR sample used in training.

38. Hosmer, D.W. and Lemeshow, S., op. cit. They point out that when the intercept is used, then there should be K-1 design or dummy variable levels in standard logistic regression. In this case shown in Fig. 3.4(b) where there is no intercept, RELR and Standard Logistic Regression under the Hosmer and Lemeshow criteria will have the same number of levels. But more generally RELR will always have K dummy variables for categorical variables with K levels.

39. Additionally, interaction effects can be formed with each specific categorical level feature. Because minimal KL divergence learning (sequential online learning) usually cannot be applied in the case of nonlongitudinal data where correlated observations cannot be assumed to be prior to one another, RELR only uses similar dummy-coded categorical features for all such correlated observations. For example, a standardized dummy code for a specific school (coded for School A versus Not School A) would be used for each different school in a model that compares high and low test scoring students. Such a categorical feature corresponding to each school will control for correlations among students within each school, as each school will have a unique regression coefficient parameter. Should a feature like a specific school be selected as a feature, then clearly this feature cannot be generalized to other schools. This specific school feature would be treated as a missing value in a more general scoring or updating sample that does not include that specific school. Even though any effect related to a specific school cannot be generalized beyond that school, at least the inclusion of an effect like that for a specific school controls for correlations in the sample that are caused by attendance at that school. And RELR's ability to handle unlimited numbers of categorical levels as standardized dummy variables even in small data samples and select unique effects for each such dummy variable greatly aids in the interpretation of such models.

40. The final selected one feature Explicit RELR model depicted in Fig. 3.4(b) was not completely nested within the final two-feature model depicted in Fig. 3.4(d) because these models had a different number of pseudo-observations. However, if we introduce a binary parameter to reflect whether or not pseudo-observations corresponding to a given feature were zeroed then this results in 1 df for this parameter and 1 df for the extra feature selected. Thus, a nested chi-square test of -2 times the difference in the *RLL* observed in an independent cross-sectional holdout sample in Fig. 3.4(c) relative

to that in Fig. 3.4(a) would be appropriate, as this is based upon the idea that -2 times the log likelihood of one model that is nested within another generates a χ^2 statistical test. This yielded a value of $-2(-42.10) - 2(-38.67) = 6.85$, which was significant ($p < 0.05$).

41. van der Laan, M.J. and Rose, S., op. cit.

42. Hastie, T., Tibshirani, R. and Friedman, J. *Elements of Statistical Learning: Data Mining, Inference, and Prediction*, (2011: Springer-Science, New York) 5th edition, online publication.

43. Ibid. Note that this definition of AIC multiples the right-hand side by N or the training sample size which is appropriate if one is only comparing relative values of AIC at a fixed training sample as in this definition.

44. Thomas Ball. Modeling First Year Achievement Using RELR, Report based upon independent research on RELR, July 2011. Can be downloaded at http://www.riceanalytics.com/_wsn/page13.html. Note that Parsed RELR was the name used at that time for what was essentially Explicit RELR without zero intercepts and perfectly balanced samples.

Chapter 4

1. Newton, I. *The Mathematical Principles of Natural Philosophy*, trans. Andrew Motte (London, 1729), pp. 387–393.

2. Padmanabhan, T. "Thermodynamical Aspects of Gravity: New insights". *Report on Progress in Physics* 73 (4) (2009): 6901; Verlinde, E.P. "On the Origin of Gravity and the Laws of Newton", *Journal of High Energy Physics* 29 (2011) 1–26.

3. Ioannidis, J.P.A. "Why Most Published Research Findings Are False", *PLoS Med* 2(8) (2005): e124 http://dx.doi.org/10.1371/journal.pmed.0020124.

4. Rosenbaum, P.R. and Rubin, D.B. "Constructing a Control Group using Multivariate Matched Sampling Incorporating the Propensity Score", *The American Statistician* 39 (1985): 33–38.

5. Imai, K. "Covariate Balancing Propensity Score", can be downloaded from Imai's Princeton University homepage.

6. Austin, P.C. "The Performance of Different Propensity-Score Methods for Estimating Differences in Proportions (Risk Differences or Absolute Risk Reductions) in Observational Studies", *Statistics in Medicine* 29(20) (2010): 2137–2148.

7. Rubin, D.B. "The Design versus the Analysis of Observational Studies for Causal Effects: Parallels with the Design of Randomized Trials", *Statistics in Medicine* 26 (1) (2007): 20–36.

8. Pearl, J. "Remarks on the Method of the Propensity Score", *Statistics in Medicine* 28 (2009): 1415–1424.

9. Gelman, A. "Resolving Disputes between Pearl, J. and Rubin, D. on Causal Inference", http://andrewgelman.com/2009/07/disputes_about/. Note that Pearl is actually simply rephrasing a comment by another participant Phil in the first part of this quote before he agrees that it is the "crux of the matter".

10. Ibid.

11. Frölich, M. "Propensity Score Matching without Conditional Independence Assumption—with an Application to the Gender Wage Gap in the United Kingdom", *Econometrics Journal* 10 (2007): 359–407.

12. Shadish, W.R. and Clark, M.H. Unpublished work reviewed by Rubin, D.B., (2007): op. cit.

13. Rose and van der Laan propose a method for causal learning that is not a propensity score method. Instead, they employ a two stage process where the initial stage is what they call a Super Learner designed to build a best predictive model based upon a large number of candidates as mentioned in the last chapter. Their Super Learner process could be an

ensemble modeling process that averages separate candidate models, or it could be a process that just returns one parsimonious model if a good parsimonious model is supplied as a candidate model by a user. In any case, as reviewed in the last chapter, this Super Learner process does require a user to specify a loss function that defines a best model. The goal of this best model is to provide an initial model that controls for all covariates in an assessment of a causal treatment effect. The second stage of this process is what Rose and van der Laan term as Targeted Maximum Likelihood Estimation or TMLE. They describe TMLE as a semiparametric method that provides an optimal tradeoff between bias and variance in the estimation of targeted effects. For binary outcomes, this TMLE updates the original Super Learner model in a logistic regression step that views the original Super Learner's model as an intercept offset and a given targeted variable as the independent variable. In this way, the TMLE is targeted toward the parameter of interest to return an estimate that has relatively low bias compared to a straight logistic regression with no such controls. The relatively unbiased targeted effect that is measured in TMLE may be interpreted as a causal factor when causal assumptions are made. Like TMLE, RELR's approach to causal reasoning completely avoids the construction of a first stage propensity score model. Yet, TMLE may be designed to test preexisting causal hypotheses rather than to discover new causal hypotheses, whereas RELR's causal reasoning process is designed to do both as it employs Explicit RELR to discover new putative causal features. There are other aspects of RELR's approach to causal effect estimates that are similar only in a very superficial way to how TMLE estimates effects in binary outcomes, as RELR also makes use of offset estimates. Yet, RELR uses the offset to adjust the covariate weights rather than the putative causal effects, and another critical distinction is that RELR is used instead of standard logistic regression in the offset regression. Because TMLE is based upon standard maximum likelihood logistic regression in its assessment of targeted effects, it may return causal effects that are likely to be more biased than RELR.

14. Rubin, D.B. "Using Multivariate Matched Sampling and Regression Adjustment to Control Bias in Observational Studies", *Journal of the American Statistical Association* 74 (1979): 318–328.

15. Other hypotheses can be tested other than concerning Explicit RELR's selected features and whether matched subgroups dictated by whether they are above or below the mean of these features differ. However, these hypothesis-testing decisions need to be made in advance of viewing the data used in the test. Obviously when hypotheses become a part of the process, there is a greater possibility of subjective biases which would be avoided by the automated procedures outlined in Appendix A8.

16. Austin, P.C. "A Critical Appraisal of Propensity Score Matching in the Medical Literature between 1996 and 2003", *Statistics in Medicine* 27(12) (2008): 2037–2049.

17. Deza, M.M. and Deza, E. *Encyclopedia of Distances*, (2013: Berlin, Springer-Verlag 2nd Edition).

18. Topsøe, F. "Some Inequalities for Information Divergence and Related Measures of Discrimination", *IEEE Transactions on Information Theory* 46 (2000): 1602–1609.

19. Ibid. This Topsøe distance is simply twice the better known Jensen–Shannon divergence measure when equal weight is given to each probability distribution that is matched. The Topsøe distance is not a metric, but its square root is a metric as it has the important triangular property. In this RELR causal reasoning application, the closest matching based upon the Topsøe distance occurs is exactly the same matching as that based upon the square root of the Topsøe distance, so this assures that this matching is based upon well behaved metric.

20. Eguchi, S. and Copas, J., op. cit.

21. Schneidman, E., Bialek, W. and Berry, M.J. "An Information Theoretic Approach to the Functional Classification of Neurons", *Advances in Neural Information Processing Systems* 15 (2003): 197–204.

22. See case study from STATA manual which shows that conditional logistic regression and McNemar's test produce very similar results; www.stata-press.com/manuals/stata10/clogit.pdf.

23. Pearl, J. *Causality*, (2009: Cambridge University Press, Cambridge 2nd Edition).

24. Hosmer, D.W. and Lemeshow, S., op. cit.

25. Ibid. They reference Breslow and Day (1980) as the source in the literature for this fact.

26. Mitchell, T., op. cit.

27. Ibid.

28. Gelman, A., Carlin, J.B., Stern, H.S. and Rubin, D.R. *Bayesian Data Analysis*, (2004: Boca Routon, Chapman and Hill/CRC 2nd Edition).

29. Ibid.

30. Crabbe, M. and Vandebrock, M. "Improving the Efficiency of Individualized Designs for the Mixed Logit Choice Model by Including Covariates", Online paper.

31. Orme, B. and Howell, J. "Applications of covariates within Sawtooth Software's HB/CBC program: theory and practical example", Sawtooth Software Conference Papers (2009: Sequoia, WA, Sawtooth Software).

32. Chen, D. "Sisterhood of Classifiers: A Comparative Study of Naïve Bayes and Noisy-or Networks", Online publication.

33. Ibid.

34. Twardy, C.R. and Tworb, K.B. "Causal interaction in Bayesian Networks", Online paper.

35. Chickering, D.M., Heckerman, D. and Meek, C. "Large-sample Learning of Bayesian Networks is NP-hard", *Journal of Machine Learning Research* 5 (2004): 1287–1330.

36. Pearson, K., Lee, A., Bramley-Moore, L. "Genetic (Reproductive) Selection: Inheritance of Fertility in Man", *Philosophical Translations of the Royal Statistical Society, Series A* 173 (1899): 534–539.

37. Simpson, E.H. (1951). "The Interpretation of Interaction in Contingency Tables", *Journal of the Royal Statistical Society, Series B* 13 (1951): 238–241.

38. Charig, C. R., Webb, D. R., Payne, S. R., Wickham, O. E. "Comparison of Treatment of Renal Calculi by Operative Surgery, Percutaneous Nephrolithotomy, and Extracorporeal Shock Wave Lithotripsy", *British Medical Journal* (*Clinical Research Ed.*) 292 (6524) (1986): 879–882.

39. Pearl, J. *Causality*, op. cit.

40. Rigmen, F. "Bayesian Networks with a Logistic Regression model for the Conditional Probabilities", *International Journal of Approximate Reasoning* 48(2) (2008): 659–666.

41. Chen, D., op. cit.

42. These comparisons are also problematic because these studies used the Receiver Operator Curve—area under the curve (ROC AUC) measure. This ROC AUC measure has been a widely used measure of classification accuracy, but has recently been shown to have some fundamental flaws. These flaws include that the AUC measure randomly fluctuates so that an accurate confidence interval often cannot be obtained and the validity of this AUC measure is a problem when the ROC from different models cross. These fundamental problems may often exist even at very large sample sizes. The fluctuation can be significant much like a stopwatch that is randomly varying by 1 s in a 100 m sprint, so the fluctuation can affect the outcome of comparisons. Hand, D., op. cit.

43. Gevaert, O., De Smet, F., Kirk, E., Van Calster, B., Bourne, T., Van Huffel, S., Moreau, Y., Timmerman, D., De Moor, B. and Condous, G. "Predicting the Outcome of Pregnancies of Unknown Location: Bayesian Networks with Expert Prior Information Compared to Logistic Regression", *Human Reproduction* 21(7) (2006): 1824–1831.

44. Le, Q.A., Strylewicz, G., Doctor, J.N. "Detecting Blood Laboratory Errors Using a Bayesian Network: An Evaluation on Liver Enzyme Tests", *Medical Decision Making* 31(2) (2011): 325–337.

45. Gopnik, A., Glymour, C., Sobel, D., Schulz, L.E., Kushnir, T. and Danks, D. "A Theory of Causal Learning in Children: Causal Maps and Bayes Nets", *Psychological Review* 111(1) (2004): 3–32.

Chapter 5

1. Cajal, S.R.y "Estructura de los Centros Nerviosos de las aves. Revista Trimestral de Histologia Normol y Patologica", First published in 1888 and translated in Clarke, E. and O'Malley, C.D. *The Human Brain and Spinal Cord: A Historical Study Illustrated by Writings from Antiquity to the Twentieth Century*, (1996: San Francisco, Norman Publishing 2nd revised edition).

2. Ibid.

3. Hocking, A.B. and Levy, W.B. "Computing Conditional Probabilities in a Minimal CA3 Pyramidal Neuron, *Neurocomputing* 65–66 (2005): 297–303.

4. Aosaki, T., Tsubokawa, H., Ishida, A., Watanabe, K., Graybiel, A.M. and Kimura, M. "Responses of Tonically Active Neurons in the Primate's Striatum Undergo Systematic Changes during Behavioral Sensorimotor Conditioning", *The Journal of Neuroscience* 14(6) (1994): 3969–3984. In this study involving implicit memory learning in monkeys, neural responses in basal ganglia, linked to targeted licking responses that were rewarded when auditory cues were presented, were observed within 10 min of training. These neural responses occurred after about 100 auditory cues and after much lower numbers of targeted licking responses.

5. Rutishauser, Mamelak, and Schuman (2006): *Neuron*, op. cit. This study in humans actually showed single trial learning in select medial temporal lobe neurons in old–new visual image recognition memory learning, although it took the recorded population recorded six trials on average to reach 90% discrimination accuracy in a predictive model.

6. The idea that field properties within the brain could have significant effects in the brain's cognitive function has received empirical support in recent years. As one example, see Nunez, P.L. and Srinivasan, R. "A Theoretical Basis for Standing and Traveling Brain Waves Measured with Human EEG with Implications for an Integrated Consciousness." *Clinical Neurophysiology*, 117(11) (2006 November): 2424–2435.

7. Not all neurons have graded potentials that directly produce spiking, as electrotonic neurons only produce graded potentials which may then help cause spiking in adjacent neurons through tight junction synapses.

8. Public domain image available at http://en.wikipedia.org/wiki/File:PurkinjeCell.jpg.

9. Avitan, L., Teicher, M. and Abeles, M. "EEG Generator—A Model of Potentials in a Volume Conductor", *Journal of Neurophysiology*, 102(5) (2009): 3046–3059.

10. Note that binary variables in RELR are also viewed as interval variables which reflect a probability of occurrence of either zero or one where missing values in independent variables will be mean imputed values which will be an interior point in this interval. The rationale behind mean-imputation is reviewed in the next chapter. Note that ordinal variables are also possible as either independent or dependent variables in RELR and care should be taken to ensure that these variables indicate rank when used to compute Pearson correlations in the error model and feature reduction.

11. http://commons.wikimedia.org/wiki/File:Artificial_Neuron_Scheme.png.

12. Hebb's idea is often simplified as "neurons that fire together wire together", but this may be too simple because it does not account for negative correlations in spiking between adjacent neurons in a feed forward network which would occur in the case of inhibitory synaptic connections. See Hebb, D.O. *The Organization of Behavior* (1949: New York, John Wiley & Sons). The research area known as spike time dependent plasticity is largely based upon Hebb's original ideas about neural learning.

13. The mathematical details are shown an appendix which will be based upon the JSM paper that can be downloaded from www.riceanalytics.com. See above note.

14. Izhikevich, E.M. "Simple Model of Spiking Neurons", *IEEE Transactions on Neural Networks* 14 (2003): 1569–1572.

15. Electronic versions of the figure and reproduction permissions are freely available at www.izhikevich.com.

16. Izhikevich, E.M. *Dynamical Systems in Neuroscience*, (2007: Cambridge MA, MIT Press). The properties of these coefficients are described on p. 155 in this reference, along with an example of how the sign of the b coefficient that determines the linear v term in the time derivative of u would determine whether this u variable has an amplifying or resonating effect. If we drop the linear v term and u variable altogether, this gives a quadratic integrate and fire neuron as shown on p. 80 of this reference. So, interesting dynamic variations are possible that reflect real neural dynamic properties simply by varying these parameters within the simple model.

17. Touboul, J. "Importance of the Cutoff Value in the Quadratic Adaptive Integrate-and-Fire Model", *Neural Computation* 21 (2009): 2114–2122.

18. Izhikevich, E.M. "Hybrid Spiking Models", *Philosophical Transactions of the Royal Society* 368(A) (2010): 5061–5070.

19. Touboul, J. and Brette, R. "Spiking Dynamics of Bidimensional Integrate-and-Fire Neurons", *SIAM Journal of Applied Dynamical Systems Society for Industrial and Applied Mathematics* 8(4) (2009): 1462–1506.

20. Izhikevich, *Dynamical Systems in Neuroscience*, op. cit., pp. 106–107.

21. Markram, H. "On simulating the brain—the next decisive years", Lecture given at *International Supercomputing Conference*, July 2011. Video at http://www.kurzweilai.net/henry-markram-simulating-the-brain-next-decisive-years.

22. Izhikevich, *Dynamical Systems in Neuroscience*, op. cit., pp. 292–294.

23. Gerstner, W. and Kistler, W. *Spiking Neuron Models, Single Neurons, Populations, Plasticity* (2002): London, Cambridge University Press. See section on multi-compartment modeling at http://icwww.epfl.ch/~gerstner/SPNM/node29.html.

24. Izhikevich, E.M. and Edelman, G.M. "Large-Scale Model of Mammalian Thalamo-cortical Systems", *PNAS* 105(9) (2008): 3593–3598.

25. In the Izhikevich simple neural dynamics model, depending upon the signal strength in a given dendritic compartment, the signals that are passed on to the next compartment can be either passive and degrade with distance or active and maintain strength with distance as a dendritic spike. The output of each dendritic compartment's signal is then passed to the next compartment in a trajectory leading to the axonal hillock.

26. London, M. and Hausser, M. "Dendritic Computation", *Annual Review of Neuroscience* 28 (2005): 503–532.

27. Gerstner and Kistler, *Spiking Neuron Models, Single Neurons, Populations, Plasticity* (2002): op. cit.

28. Izhikevich and Edelman, *PNAS* (2008): op. cit. See description of synaptic dynamics and plasticity in their appendix.

29. Schneidman, E., Berry, M.J., Segev, R. and Bialek, W. "Weak Pairwise Correlations Imply Strongly Correlated Network States in a Neural Population", *Nature* 440(20) (2006): 1007–1012.

30. RELR allows very specific higher order nonlinear or interaction effects to be selected without selecting lower order effects. In other words, it is not subject to the principle of marginality suggested by the late John Nelder which would require for example that two main effect terms always be selected whenever an interaction between these two effects is selected. Nelder pointed out that arbitrary scaling of a variable relative to others will arbitrarily influence the regression model if the marginality is not honored as in the selection of main effects when an interaction is selected. The reason that RELR is not

subject to this limitation is that all independent variables, including all interaction and nonlinear effects, are standardized to become z-scores with a mean of 0 and a standard deviation of 1. Additionally each interaction is computed based upon standardized variables, so whether temperature is initially measured in Celsius or Fahrenheit units has no influence on the RELR model. See Nelder, *Journal of Applied Statistics* (2000): op. cit.

31. Rutishauser, Mamelak, and Schuman (2006): *Neuron*, op. cit.
32. Xu, F. and Tenenbaum, J.B. "Word Learning as Bayesian Inference", *Psychological Review* 114 (2007): 245–272.
33. Xu, F. and Tenenbaum, J.B. "Sensitivity to Sampling in Bayesian Word Learning", *Developmental Science* 10 (2007): 288–297.
34. Izhikevich, *Dynamical Systems in Neuroscience*, op. cit., p. 2.
35. If we assume 1000 independent linear inputs with two way interactions and no nonlinear effects, then we get $(1000^2/2) + 1000$ or 500,000 two way interactions and 5,001,000 total independent variables.
36. We simply multiply our 500,000 two way interactions by four and we add 4×1000 effects for the linear main effects and the associated quadratic, cubic, and quartic effects to give 2,004,000 effects as candidate effects.
37. Dahlquist, G. and Björk, Å. *Numerical Methods*, (1974: Englewood Cliffs, Prentice Hall). See pp. 101–103 on the Runge Phenomenon. This Dahlquist and Björk rule is that the degree of polynomial n must be less than $2\sqrt{m}$, where m equals the number of sampled points, in a polynomial interpolation system of equations. In a multivariate regression model with many linear main effects, polynomial effects, and interaction effects, then n may be considered to refer to the total number of independent variable effects and m to the number of training sample observations. This is because the number of variables in a system of polynomial interpolating equations also includes all interaction terms and nonlinear terms. Note that when there are a small number of linear variables with no interaction or nonlinear terms as traditionally occurred in standard regression methods before the era of Big Data due to processing speed constraints, this rule is basically the traditional "10:1 rule" that was used to determine how many independent variable effects could be employed in a regression model given the number the observations. For example, if we have 400 observations or $m = 400$ and if we assume that each independent variable is like a separate polynomial effect in an interpolating system of equations, then the Dahlquist and Björk rule tells me that I can choose 40 linear main effect variables or $n = 40$ in this case, which is exactly the tradition 10:1 rule that we must have 10 observations for every variable.
38. Ibid. Assuming that a neuron fires 100 times a second, we get 360,000 spike responses in an hour. Assuming that we employ Dahlquist and Björk rule and call m, the number of spike responses, and N, the number of total features including interaction effects, then $m = N^2/2$ or $m = 501,000^2/2$, which implies that $m = 125,500,500,000$ spikes. So at the rate of 360,000 spikes/h, we have 125,500,500,000 spikes/360,000 spikes/h or about 348,612 h or about 40 years for even minimally reliable learning.
39. Random Forest methods suffer from their own problems that have been highlighted in recent years such as biased solutions that are still corrupted by multicollinearity and other variable selection problems. See Strobl, C. et al. (2007) op. cit.
40. The instability of L2 norm Penalized Logistic Regression parameters with small samples is exemplified in a model reviewed in Rice, D.M. "Generalized Reduced Error Logistic Regression Machine", *Section on Statistical Computing—JSM Proceedings* (2008): 3855–3862.
41. Poggio, T. and Girosi, F. "Extensions of a Theory of Networks for Approximation and Learning: Dimensionality Reduction and Clustering", *AI Memo/CBIP Paper* 1167(44) (1990): 1–18.
42. Mitchell, T.M. "Generative and Discriminative Classifiers: Naive Bayes and Logistic Regression", Chapter 1 in *Machine Learning*, Online book (2010): 1–17.

43. Hebb, D.O. (1949). The organization of behavior. New York: Wiley & Sons.
44. Rosenblatt, F. "The Perceptron—A Perceiving and Recognizing Automaton", *Report 85-460-1*, Cornell Aeronautical Laboratory (1957). The perceptron performed classification but not regression, so the perceptron is still quite distinct from today's standard maximum likelihood regression.
45. Minsky, M.L. and Papert, S.A. *Perceptrons* (1969: Cambridge MA, MIT Press).
46. Rumelhart, D.E., Hinton, G.E. and Williams, R. J. "Learning Internal Representations by Error Propagation", In Rumelhart, D.E., McClelland, J.L. and the PDP research group. (eds), *Parallel Distributed Processing: Explorations in the Microstructure of Cognition, Volume 1: Foundations* (1986: Cambridge MA, MIT Press) 318–362.
47. Clopath, C. and Gerstner, W. "Voltage and Spike-Timing Interact in STDP – A Unified Model", *Frontiers in Synaptic Neuroscience* 2 (2010) 1–11.
48. Poirazi, P., Brannon, T. and Mel, B.W. "Pyramidal Neuron as a 2-Layer Neural Network", *Neuron* 37 (2003): 989–999.
49. London and Hausser, *Annual Review of Neuroscience*, (2005): op. cit.
50. Barra, A., Bernacchia, A., Santucci, E., Contucci, P. "On the equivalence of Hopfield Networks and Boltzmann Machines", *Neural Networks* 34 (2012): 1–9.
51. Hopfield, J.J. "Neural Networks and Physical Systems with Emergent Collective Computational Abilities", *Proceedings of the National Academy of Sciences of the United States of America* 79 (1982): 2554–2558.
52. Hinton, G.E. "Boltzmann Machine", *Scholarpedia* 2(5) (2007): 1668. Note that IBM's current neuromorphic cognitive architecture is based in part on Restricted Boltzmann Machines. See https://dl.dropboxusercontent.com/u/91714474/Papers/021.IJCNN2013. Applications.pdf.
53. Hopfield, J.J. "Hopfield Network", *Scholarpedia* 2(5) (2007): 1977. Associative memory networks based upon Hopfield networks or Restricted Boltzmann Machines seem to have been applied in engineering much more than in business or medicine to date, which is similar to other artificial neural network methods. But as reviewed earlier the Netflix model that was actually used was a simple ensemble of a Restricted Boltzmann Machine and an SVM model; *Netflix Tech Blog* op cit.
54. Hinton, G.E., op. cit.
55. Liou, C.Y. and Yuan, S.K. "Error Tolerant Associative Memory", *Biological Cybernetics* 81 (1999): 331–342. Another significant development beyond what occurred in the 1980's research is that Hopfield associative memory models do lend themselves to real-time hardware configurations using memristors. See Farnood Merrikh-Bayat and Saeed Bagheri-Shouraki, Efficient neuro-fuzzy system and its Memristor Crossbar-based Hardware Implementation, http://arxiv.org/pdf/1103.1156v1.pdf. Other advances developed by Saffron Technologies now appear to avoid the pitfalls related to physical scaling limitations of the crossbar memristor implementations (Personal Communication, Manny Aparicio Co-founder and Paul Hofmann CTO, Saffron Technology, Cary, NC).
56. Storkey, A. "Increasing the Capacity of a Hopfield Network without Sacrificing Functionality", *Artificial Neural Networks—ICANN'97* (1997): 451–456. However, some increase in complexity with the memristors referenced in the previous note might be a major improvement with huge capacity increases. Yet, associative memory networks are still limited in that they only work with binary features, so human judgment is still involved in producing binary features from non-binary variables in real world applications, as this apparently cannot be done as an automatic associative learning process at present. So it is not yet a hands-free approach to machine learning.
57. Riesenhuber, M. and Poggio, T. "A New Framework of Object Categorization in Cortex" Biologically *Motivated Computer Vision, First IEEE International Workshop Proceedings*, (2000): 1–9.

58. The unsupervised K-means clustering approach used in Poggio's group responds with binary signals that prefer visual input data along one dimension, such as the orientation of lines in a specific direction in the visual field. This selective tuning to one dimension of information is similar to how simple cells in the primary visual cortex respond. In these Poggio simulations, later higher level visual processing that models inferotemporal cortex object categorization incorporates such features as inputs and classifies the objects using classic supervised neural networks.

59. Kohonen, T. "Self-Organized Formation of Topologically Correct Feature Maps", *Biological Cybernetics* 43 (1982): 59–69.

60. Izhikevich, E.M. "Polychronization: Computation with Spikes", *Neural Computing* 18 (2006):245–282.

61. Edelman, G. *Neural Darwinism: The Theory of Neuronal Group Selection* (1987: New York, Basic Books).

62. Calvin, W.H. *The Cerebral Code* (1996: Cambridge, MIT Press).

63. Izhikevich, E.M. and Edelman, G.M. "Large-Scale Model of Mammalian Thalamo-cortical Systems", *PNAS* 105 (2008): 3593–3598.

64. Another approach to circumvent problems with the standard supervised/unsupervised learning paradigm is the semi-supervised learning approach which initially seeds a set of responses with human generated labels that determine supervised learning and then proceeds with unsupervised learning from this initial set. See Mitchell, "Semi-Supervised Learning over Text" (2006): op. cit. However, semi-supervised learning suffers from requiring humans to label responses, so it is obviously not an automated machine learning solution. Under specific seeding conditions, semi-supervised learning is argued to be a model for reinforcement learning. See Damljanovi, D.D. *Natural Language Interfaces to Conceptual Models*, Ph.D. thesis. The University of Shefeld Department of Computer Science July 2011. Also see Sutton, R.S. and Barto, A.G. *Reinforcement Learning: an Introduction* (MIT Press, Cambridge, MA, 1998) for a more general discussion of machine reinforcement learning. Note that reinforcement learning also can be modeled as a time varying hazard model similar to survival analysis. See Alexander, W.H. and Brown, J.W., "Hyperbolically Discounted Temporal Difference Learning", *Neural Computation* 22(6) (2010):1511–1527.

65. Van Belle, T. "Is Neural Darwinism Darwinism?" *Artificial Life* 3(1) (1997): 41–49.

66. Fernando, C., Karishma, K.K. and Szathmáry, E. "Copying and Evolution of Neuronal Topology", *PLoS One* 3(11) (2008): e3775. http://dx.doi.org/10.1371/journal.pone.0003775.

67. Izhikevich, E.M. and Edelman, G.M. "Large-Scale Model of Mammalian Thalamo-cortical Systems", *PNAS* 105 (2008): 3593–3598.

68. Furber, S.B., Lester, D.R., Plana, L.A., Garside, J.D., Painkras, E., Temple, S. and Brown, A.D. "Overview of the SpiNNaker System Architecture Computers", *IEEE Transactions on Computers* PP(99) (2012) 1.

69. Hodgkin and Huxley, *Journal of Physiology* (1952): op. cit.

70. John, E.R., op. cit.

71. Nunez, P.L. and Srinivasan, R., op. cit.

72. Anastassiou, C.A., Perin, R., Markram, H., Koch, C. "Ephaptic Coupling of Cortical Neurons", *Nature Neuroscience* 14(2) (2011): 217.

Chapter 6

1. Schopenhauer, A. *The World as Will and Representation*, translated by Payne, E.F.J. (1958: Indian Hills, Colorado, The Falcon's Wing).

2. Alivisatos, A.P., Chun, M., Church, G.M., Greenspan, R.J., Roukes, M.L. and Yuste, R. "The Brain Activity Map Project and the Challenge of Functional Connectomics", *Neuron* 74(6) (2012): 970–974.

3. John, E.R. "The Neurophysics of Consciousness", *Brain Research Reviews* 39 (2002): 1–28.

4. Ibid.
5. Ibid.
6. The scalp EEG is not only much noisier, but is also most sensitive to field potentials that arise from current oriented in the direction of electrodes in apical pyramidal neurons in cerebral cortex neurons. Pizzagalli, D.E. "Electroencephalography and High-Density Electrophysiological Source Localization", Chapter prepared for Cacioppo, J.T. et al., *Handbook of Psychophysiology* (3rd Edition) online preprint; the MEG on the other hand is most sensitive to current that exists in a tangential direction with respect to the surface of the scalp.
7. Penfield, W. and Erickson, T.C. Epilepsy and Cerebral Localization: A Study of the Mechanism, Treatment and Prevention of Epileptic Seizures (1941: Springfield IL, Thomas, C.C.).
8. Rice, D.M. "The EEG and the Human Hypnagogic Sleep State", Master's Empirical Research Thesis, (1983: Durham NH, University of New Hampshire).
9. Schacter, D.L. "The Hypnagogic State: A Critical Review of the Literature", *Psychological Bulletin* 83 (1976): 452–481.
10. Tonini, G. "Consciousness as Integrated Information: A Provisional Manifesto", *Biological Bulletin* 215 (2008): 216–242.
11. Rice, D.M. The EEG and the Human Hypnagogic Sleep State, Master's Empirical Research Thesis, (1983: Durham NH, University of New Hampshire).
12. https://commons.wikimedia.org/wiki/File:Brain_headBorder.jpg.
13. Nunez, P.L. and Srinivasan, R., op. cit.
14. Hameroff, S. "The 'Conscious Pilot'—Dendritic Synchrony Moves through the Brain to Mediate Consciousness", *Journal of Biological Physics* 36 (1) (2009): 71–93.
15. Anderson, C.H. and DeAngelis, G.C. "Redundant Populations of Simple Cells Represent Wavelet Coefficients in Monkey V1", *Journal of Vision* 5(8) (2005): article 669.
16. Hubel, D.H. and Wiesel, T.N. "Receptive Fields, Binocular Interaction and Functional Architecture in the Cat's Visual Cortex", *Journal of Physiology* 160 (1962): 106–154.
17. Berens, P., Ecker, A.S., Cotton, R.J., Ma, W.J., Bethge, M., and Tolias, A.S. "A Fast and Simple Population Code for Orientation in Primate V1", *The Journal of Neuroscience* 32(31) (2012): 10618–10626.
18. Gross, C.G. "Genealogy of the Grandmother Cell", *Neuroscientist* 8(5): (2002): 512–518.
19. Connor, C. "Friends and grandmothers", *Nature* 435 (7045) (2005): 1036–1037.
20. Original image in Lesher, G.W. and Mingolla, E. "The Role of Edges and Line-ends in Illusory Contour Formation", *Vision Research* 33(16) (1993): 2263–2270; Redrawn with permission from Elsevier.
21. Wertheimer, M., *Laws of Organization in Perceptual Forms*, First published as Untersuchungen zur Lehre von der Gestalt II, in *Psycologische Forschung*, 4 (1923): 301–350. Translation published in Ellis, W., *A Source Book of Gestalt Psychology* (1938: London, Routledge and Kegan Paul).
22. Singer, W. "Neuronal Synchrony: A Versatile Code for the Definition of Relations?", *Neuron* 24 (1999): 49–65.
23. Singer, W. "Binding by Synchrony", *Scholarpedia* 2(12) (2007): 1657.
24. Ibid.
25. Gray, C.M., Konig, P., Engel, A.K., Singer, W. "Oscillatory Responses in Cat Visual Cortex Exhibit Inter-Columnar Synchronization which Reflects Global Stimulus Properties", *Nature* 338 (1989): 334–337.
26. John, E.R., op. cit.
27. Anastassiou, C.A., Perin, R., Markram, H., Koch, C., op. cit.
28. Werning, M. and Maye, A. "The Cortical Implementation of Complex Attribute and Substance Concepts Synchrony, Frames, and Hierarchical Binding", *Chaos and Complexity Letters* 2(2/3) (2007): 435–452.

29. Engel, A.K., König, P., Kreiter, A.K. and Singer, W. "Interhemispheric Synchronization of Oscillatory Neuronal Responses in Cat Visual Cortex", *Science* 252 (1991): 1177–1179.
30. Thiele, A. and Stoner, G. "Neuronal Synchrony Does Not Correlate with Motion Coherence in Cortical Area MT", *Nature* 421 (6921) (2003): 366–370.
31. Dong, Y., Mihalas, S., Qiu, F., von der Heydt, R. and Niebur, E. "Synchrony and the Binding Problem in Macaque Visual Cortex", *Journal of Vision* 8 (7) (2008): 1–16.
32. Singer, W. Binding by synchrony, op. cit.
33. Eckhorn, R., Bruns, A., Saam, M., Gail, A., Gabriel, A. and Brinksmeyer, H.J. "Flexible Cortical Gamma–Band Correlations Suggest Neural Principles of Visual Processing", *Visual Cognition* 8(3–5) (2001): 519–530.
34. Izhikevich, E.M. and Hoppensteadt, F.C. "Polychronous Wavefront Computations", *International Journal of Bifurcation and Chaos* 19(5) (2009): 1733–1739.
35. The Werning, M. and Maye, A., op. cit. (2007) model also allows for traveling waves.
36. Hameroff, S. "The 'Conscious Pilot'—Dendritic Synchrony Moves through the Brain to Mediate Consciousness", op. cit.
37. Hebb, D.O. *The organization of behavior*, op. cit.
38. Pallas, S.L. (ed.) *Developmental Plasticity of Inhibitory Circuitry*, Introductory chapter (also by same author), pp. 3–12, (2010: New York, Springer).
39. Rutishauser, U., Ross, I.B., Mamelak, A.N., Schuman, E.M. "Human Memory Strength is Predicted by Theta-Frequency Phase-Locking of Single Neurons", *Nature* 464(7290) (2010): 903–907.
40. Sjöström, J. and Gerstner, W. "Spike Timing Dependent Plasticity", *Scholarpedia* (2010): 1362.
41. Clopath, G. and Gerstner, W. "Voltage and Spike-Timing Interact in STDP—A Unified Model", *Frontiers in Synaptic Neuroscience* 2 (2010): 1–11.
42. Ibid. The previous references are a good starting point for research on the fine details of this mechanism along with current reasonable speculations.
43. Gaba synapses may start out by being excitatory but may become inhibitory over the course of development. See Pallas, S.L. (ed.) Developmental Plasticity of Inhibitory Circuitry, op. cit.
44. Rice, D.M. and Hagstrom, E.C. "Some Evidence in Support of a Relationship between Human Auditory Signal-Detection Performance and the Phase of the Alpha Cycle", op. cit.
45. Busch, N.A. and VanRullen, R. "Spontaneous EEG Oscillations Reveal Periodic Sampling of Visual Attention", *PNAS* 107(7) (2010): 16038–16043.
46. Desimone, R. "Neural Synchrony and Selective Attention", Lecture given at Boston University on Feb. 27, 2008. http://www.youtube.com/watch?v=GdDMzV26WSk.
47. Desimone, R., op. cit.
48. Buzsáki, G. Rhythms of the Brain, op. cit.
49. Axmacher, N., Henseler, M.M., Jensend, O., Weinreich, I., Elger, C.E. and Fell, J. "Cross-Frequency Coupling Supports Multi-Item Working Memory in the Human Hippocampus", *PNAS* 107(7) (2010): 3228–3233.
50. Fell, J. and Axmacher, N. "The Role of Phase Synchronization in Memory Processes", *Nature Reviews Neuroscience* 12 (2011): 105–118.
51. Kamiński, J., Brzezicka, A. and Wróbel, A. "Short-Term Memory Capacity (7 ± 2) Predicted by Theta to Gamma Cycle Length Ratio", *Neurobiology of Learning and Memory* 95(1) 2011: 19–23.
52. Lisman, J.E. and Idiart, M.A.P. "Storage of 7 + 2 Short-Term Memories in Oscillatory Subcycles", *Science* 267 (1995): 1512–1514.
53. Fell, J. and Axmacher, N. "The Role of Phase Synchronization in Memory Processes", op. cit.
54. Craik, F.I.M. and Tulving, E., op. cit.

55. Carlqvist, H, Nikulin, V.V., Strömberg, J.O. and Brismar, T. "Amplitude and Phase Relationship between Alpha and Beta Oscillations in the Human Electroencephalogram", *Medical and Biological Engineering and Computing* 43(5) (2005): 599–607.

56. Nikulin, V.V. and Brismar, T. "Phase Synchronization between Alpha and Beta Oscillations in the Human Electroencephalogram", *Neuroscience* 137(2) 2006: 647–657.

57. Palva, J.P., Palva, S. and Kaila, K. "Phase Synchrony among Neuronal Oscillations in the Human Cortex", *The Journal of Neuroscience* 2005 25(15) (2005): 3962–3972.

58. Roopun, A.K., Kramer, M.A., Carracedo, L.M., Kaiser, M., Davies, C.H., Traub, R.D., Kopell, N.J. and Whittington, M. "Temporal Interactions between Cortical Rhythms", *Frontiers in Neuroscience* 2(2) (2008): 145–154.

59. Sauseng, P., Klimesch, W., Gruber, W.R., Birmbaumer, N. "Cross-Frequency Phase Synchronisation: A Brain Mechanism of Memory Matching and Attention", *Neuroimage* 40 (2008): 308–317.

60. Gaona, C.M., Sharma, M., Freudenburg, Z.V., Breshears, J.D., Bundy, D.T., Roland, J., Barbour,D.L., Schalk, G. and Leuthardt, E.C. "Nonuniform High-Gamma (60–500 Hz) Power Changes Dissociate Cognitive Task and Anatomy in Human Cortex", *The Journal of Neuroscience* 31(6) (2011): 2091–2100.

61. Jacobs, J. and Kahana, M.J. "Neural Representations of Individual Stimuli in Humans Revealed by Gamma–Band Electrocorticographic Activity", *The Journal of Neuroscience* 29(33) (2009): 10203–10214.

62. Canolty, R.T., Edwards, E., Dalal, S.S., Soltani, M., Nagarajan, S.S., Kirsch, H.E., Berger, M.S., Barbaro, N.M. and Knight, R.T. "High Gamma Power is Phase-Locked to Theta Oscillations in Human Neocortex", *Science* 313(5793) (2006): 1626–1628.

63. Voytek, B., Canolty, R.T., Shestyuk, A., Crone, N.E., Parvizi, J. and Knight, R.T. "Shifts in Gamma Phase–Amplitude Coupling Frequency from Theta to Alpha over Posterior Cortex during Visual Tasks", *Frontiers in Human Neuroscience* 4:191 (2010): http://dx.doi.org/10.3389/fnhum.2010.00191.

64. Kassam, K.S., Markey, A.R., Cherkassky,V.L., Loewenstein, G., Just, M.A. "Identifying Emotions on the Basis of Neural Activation", *PLoS One* 8(6) (2013): e66032. http://dx.doi.org/10.1371/journal.pone.0066032.

Chapter 7

1. Published by Vigyan Prasar, New Delhi, India.

2. Abitz, M., Nielsen, R.D., Jones, E.G., Laursen, H., Graem, N., Pakkenberg, B. "Excess of Neurons in the Human Newborn Mediodorsal Thalamus Compared with that of the Adult", *Cerebral Cortex* 17 (11) (2007): 2573–2578.

3. Huttenlocher, P.R. "Synaptic Plasticity and Elimination in Developing Cerebral Cortex", *Journal of Mental Deficiency* 88(5) (1984): 488–496.

4. Petanjek, Z., Judaš, M., Šimić, G., Rašin, M.R., Uylings, H.B.M., Rakic, P. and Kostović, I. "Extraordinary Neoteny of Synaptic Spines in the Human Prefrontal Cortex", *PNAS* 2011 108(32) (2011): 13281–13286.

5. Changeux, J.P. *Neuronal Man* (Laurence Garey, translator) (1985: New York: Pantheon Books).

6. Black, J.E. and Greenough, W.T. "Induction of Pattern in Neural Structure by Experience: Implications for Cognitive development", In Lamb, M.E., Brown, A.L. & Rogott, B. (eds), *Advances in Developmental Psychology* Vol. 4 (pp. 1–50). (1986: Hillsdale NJ, Erlbaum).

7. Fu, M., Yu, X., Lu, J., Zuo, Y. "Repetitive Motor Learning Induces Coordinated Formation of Clustered Dendritic Spines in vivo", *Nature* 483 (2012): 92–95.

8. Uhlhaas, P.J., Roux, F., Rodriguez, E., Rotarska-Jagiela, A. and Singer, W. "Neural Synchrony and the Development of Cortical Networks", *Trends in Cognitive Sciences* 14(2) (2010): 72–80.

9. Shaw, P., Greenstein, D., Lerch, J., Clasen, L., Lenroot, R., Gogtay, N., Evans, A., Rapoport, J., Giedd, J. "Intellectual Ability and Cortical Development in Children and Adolescents", *Nature* 440 (2006): 676–679.
10. Ibid.
11. Choi, Y.Y., Shamosh, N.A., Cho, S.H., DeYoung, C.G., Lee, M.J., Lee, J.M., Kim, S.I., Cho, Z.H., Kim, K., Gray, J.R. and Lee, K.H. "Multiple Bases of Human Intelligence Revealed by Cortical Thickness and Neural Activation", *The Journal of Neuroscience* 28(41) (2008):10323–10329.
12. Sowell, E.R., Peterson, B.S., Thompson, P.M., Welcome, S.E., Henkenius, A.L., Toga, A.W. "Mapping Cortical Change across the Human Life Span", *Nature Neuroscience* 6 (2003): 309–315.
13. Burke, S.N. and Barnes, C.A. "Neural Plasticity in the Ageing Brain", *Nature Reviews Neuroscience* 7 (2006): 30–40. They report that one area where substantial age-related neuronal loss is observed is area 8a of dorsolateral prefrontal cortex in nonhuman primates where a 30% age-related neuronal loss is associated with working memory deficits.
14. Hof, P.R. and Morrison, J.H. "The Aging Brain: Morphomolecular Senescence of Cortical Circuits", *Trends in Neurosciences* 27(10) (2004): 607–613.
15. Burke, S.N. and Barnes, C.A., op. cit.
16. Duan, H., Wearne, S.L., Rocher, A.B., Macedo, A., Morrison, J.H. and Hof, P.R. "Age-related Dendritic and Spine Changes in Corticocortically Projecting Neurons in Macaque Monkeys", *Cerebral Cortex* 13 (2003): 950–961.
17. Fernández, A., Arrazola, J., Maestú, F., Amo, C., Gil-Gregorio, P., Wienbruch, C. and Ortiz, T. "Correlations of Hippocampal Atrophy and Focal Low-Frequency Magnetic Activity in Alzheimer Disease: Volumetric MR Imaging-Magnetoencephalographic Study", *American Journal of Neuroradiology* 24(3) (2003): 481–487.
18. Grunwald, M., Hensel, A., Wolf, H., Weiss, T. and Gertz, H.G. "Does the Hippocampal Atrophy Correlate with the Cortical Theta Power in Elderly Subjects with a Range of Cognitive Impairment?" *Journal of Clinical Neurophysiology* 24(1) (2007): 22–26.
19. Breslau, J., Starr, A., Sicotte, N., Higa, J. and Buchsbaum, M.S. "Topographic EEG Changes with Normal Aging and SDAT", *Electroencephalography and Clinical Neurophysiology* 72(4) (1989): 281–289.
20. Rice, D.M. et al. *Journals of Gerontology: Medical Sciences* (1990): op. cit.
21. Visser, S.L., Hooijer, C., Jonker, C., Van Tilburg, W. and De Rijke, W. "Anterior Temporal Focal Abnormalities in EEG in Normal Aged Subjects; Correlations with Psychopathological and CT Brain Scan Findings", *Electroencephalography and Clinical Neurophysiology* 66(1) (1987): 1–7.
22. Rice, D.M. et al. *Journals of Gerontology: Psychological Sciences* (1991): op. cit.
23. Boon, M.E., Melis, R.J.F., Rikkert, M.G.M.O., Kessels, R.P.C. "Atrophy in the Medial Temporal Lobe is specifically Associated with Encoding and Storage of Verbal Information in MCI and Alzheimer Patients", *Journal of Neurology Research* 1(1) (2011): 11–15.
24. Henneman, W.J.P., Sluimer, J.D., Barnes, J., van der Flier, W.M., Sluimer, I.C., Fox, N.C., Scheltens, P., Vrenken, H. and Barkhof, F. "Hippocampal Atrophy Rates in Alzheimer Disease: Added Value over Whole Brain Volume Measures", *Neurology* 72 (2009): 999–1007.
25. Ibid.
26. Duara, R., Loewenstein, D.A., Potter, E., Appel, J., Greig, M.T., Urs, R., Shen, Q., Raj, A., Small, B., Barker, W., Schofield, E., Wu, Y. and Potter, H. "Medial Temporal Lobe Atrophy on MRI Scans and the Diagnosis of Alzheimer Disease", *Neurology* 71(24) (2008): 1986–1992.

27. Mosconi, L. "Brain Glucose Metabolism in the Early and Specific Diagnosis of Alzheimer's Disease: FDG-PET studies in MCI and AD", *European Journal of Nuclear Medicine and Molecular Imaging* 32(4) (2005): 486–510.

28. At present, efforts to use PET scans to image the putative causal beta amyloid protein in living patients have not yet received FDA approval.

29. Koivunen, J., Scheinin, N., Virta, J.R., Aalto, S., Vahlberg, T., Någren, K., Helin, S., Parkkola, R., Viitanen, M. and Rinne, J.O. "Amyloid PET Imaging in Patients with Mild Cognitive Impairment: A 2-year Follow-up Study", *Neurology* 76(12) (2011): 1085–1090.

30. Marchione, M., op. cit.

31. Wechsler Memory Scale, Revised. Now it is Wechsler Memory Scale-IV, marketed by Pearson, San Antonio, TX.

32. Stories like this one have been part of the Wechsler Memory Scale for many years. This particular story comes from a talk given at the Washington University Alzheimer's Center in St Louis by Albert, M.S., http://www.alz.washington.edu/NONMEMBER/SPR10/Albert.pdf.

33. This simulation gives almost complete separation—as there is no almost no error other than rounding error. In a case with substantial noise, the initial episodic learning often may show greater magnitudes of weights with smaller episodic samples—and those weights may decrease somewhat with larger episodic samples as demonstrated in Fig. 3.1. Under the RELR model, this leads to an interpretation that a larger magnitude of positive or negative weights in novel learning episodes will reflect overfitting if that novel learning episode is not representative of future data. These learning-related changes in RELR regression coefficient weight magnitudes can be interpreted as a model of long term potentiation (LTP) and long term depression (LDP). Rice, D.M. "Simulated Learning and Memory Effects in RELR", *Society for Neuroscience 2013 Conference Abstracts*, in press.

34. Howieson, D.B., Mattek, N., Seeyle, A.M., Dodge, H.H., Wasserman, D., Zitzelberger, T. and Kaye, J.A. "Serial Position Effects in Mild Cognitive Impairment", *Journal of Clinical and Experimental Neuropsychology* 33(3) (2011): 292–299.

35. Ibid. Note that the point that is made in the next sentence in Chapter 7 about Explicit RELR and missing values in Alzheimer's disease is corrected as best as possible in the actual real world implementations of RELR with the usage of Welch's *t* values to handle missing data in the RELR error model and feature reduction as explained in the appendix. But obviously, the brain does not have good mechanisms to correct the susceptibility its explicit memory to missing information due to neural loss in Alzheimer's disease.

36. Lehn, H., Steffenach, H.A., van Strien, N.M., Veltman, D.J., Witter, M.P. and Håberg, A.K. "A Specific Role of the Human Hippocampus in Recall of Temporal Sequences", *The Journal of Neuroscience* 29(11) (2009): 3475–3484.

37. Kesner, R.P., Gilbert, P.E., Barua, L.A. "The Role of the Hippocampus in Memory for the Temporal Order of a Sequence of Odors", *Behavioral Neuroscience* 116(2) (2002): 286–290.

38. Farovik, A., Dupont, L.M., Eichenbaum, H. "Distinct Roles for Dorsal CA3 and CA1 in Memory for Sequential Non-spatial Events", *Learning and Memory* 17(1) (2009): 12–17.

39. Thorvaldsson, V., MacDonald, S.W.S., Fratiglioni, L., Winblad, B., Kivipelto, M., Laukka, E.J., Skoog, I., Sacuiu, S., Guo, X., Östling, S., Börjesson-Hanson, A., Gustafson, D., Johansson, B. and Bäckman, L. "Onset and Rate of Cognitive Change before Dementia Diagnosis: Findings from Two Swedish Population-Based Longitudinal Studies", *Journal of the International Neuropsychological Society* 17(1) (2011): 154–162.

40. Rice et al. *Journals of Gerontology* (1991), op. cit.

41. Anderson, K.L., Rajagovindan, R., Ghacibeh, G.A., Meador, K.J. and Ding, M. "Theta Oscillations Mediate Interaction between Prefrontal Cortex and Medial Temporal Lobe in Human Memory", *Cerebral Cortex* 20 (7) (2010): 1604–1612.

42. Rutishauser, U., Ross, I.B., Mamelak, A.N., Schuman, E.M. "Human Memory Strength is Predicted by Theta-Frequency Phase-Locking of Single Neurons", op. cit.

43. Greenwood, P.M. and Parasuraman, R. "Neuronal and Cognitive Plasticity: A Neurocognitive Framework for Ameliorating Cognitive Aging", *Frontiers in Aging Neuroscience* 2:150 (2010).

44. Ibid.

45. Milgram, N.W., Head, E., Muggenburg, B., Holowachuk, D., Murphey, H., Estrada, J., Ikeda-Douglas, C.J., Zicker, S.C. and Cotman, C.W. (2002). "Landmark Discrimination Learning in the Dog: Effects of Age, an Antioxidant Fortified Food, and Cognitive Strategy", *Neuroscience Biobehavioral Reviews* 26 (2002): 679–695. While these are very interesting results and an excellent study, this study might not have been able to control exercise independently from the cognitive stimulation treatment through the inclusion of walking.

46. Fabre, C., Chamari, K., Mucci, P., Massé-Biron, J., Préfaut, C. "Improvement of Cognitive Function by Mental and/or Individualized Aerobic Training in Healthy Elderly Subjects", *International Journal of Sports Medicine* 23(6) (2002): 415–421.

47. Owen, A.M., Hampshire, A., Grahn, J.A., Stenton, R., Dajan, S., Burns, A.S., Howard, R.J. and Ballard, C.G. "Putting Brain Training to the Test", *Nature* 465 (2010): 775–779.

48. Borness, C., Proudfoot, J., Crawford, J., Valenzuela, M. "Putting Brain Training to the Test in the Workplace: A Randomized, Blinded, Multisite, Active-Controlled Trial", *PLoS One* 8(3) (2013): e59982. http://dx.doi.org/10.1371/journal.pone.0059982.

49. Descartes and the Pineal Gland. (Stanford Encyclopedia of Philosophy).

50. Anastassiou, C.A., Perin, R., Markram, H., Koch, C., op. cit.

51. Edelman, G.M. "Naturalizing Consciousness: A Theoretical Framework", *PNAS* 100(9) (2003): 5520–5524.

52. Edelman, G.M., Gally, J.A. and Baars, B.J. "Biology of Consciousness", *Frontiers in Psychology* 25 January (2011): http://dx.doi.org/10.3389/fpsyg.2011.00004.

53. McFadden, J. "The CEMI Field Theory: Seven Clues to the Nature of Consciousness". In Tuszynski, J.A. *The Emerging Physics of Consciousness* pp. 385–404 (2006: Berlin: Springer).

54. Tse, P.U. *The Neural Basis of Free Will*, (Cambridge, MIT Press, 2013).

55. Hameroff, S. (October 22, 2010). "Clarifying the tubulin bit/qubit—Defending the Penrose-Hameroff Orch OR Model of Quantum Computation in Microtubules". Google Workshop on Quantum Biology. http://www.youtube.com/watch?v=LXFFbxoHp3s.

56. Hameroff, S. "The 'Conscious Pilot'—Dendritic Synchrony Moves through the Brain to Mediate Consciousness", op. cit.

57. Kurzweil, R. Your Brain in the Cloud, http://www.youtube.com/watch?v=0iTq0FLDII4, published on Feb. 19, 2013.

Chapter 8

1. Leibniz, G. 1685, *The Art of Discovery*, op. cit.

2. Mutual information is sensitive to both linear and nonlinear effects, whereas Pearson correlation is only sensitive to linear effects unless nonlinear effects are defined through polynomial terms as is done in RELR. Yet, all important models of neural dynamics reviewed in Chapter 5 are also based upon similar polynomial expressions as RELR, and it is unclear how to formulate models of neural dynamics that are capable of separable linear and nonlinear effect learning with mutual information, whereas all linear and nonlinear

effects can be separated in RELR's correlation-based learning and are thus capable of separable spike-timing dependent learning as would be required if such effects code learned features. The larger issue is the lack of agreement in how to define these information theory concepts not used in the present book. To get a sense of different operational definitions of minimum description length computation and Kolmogorov Complexity, see Ming Li and Paul Vitányi, An Introduction to Kolmogorov Complexity and its Applications, (2008, New York, Springer-Verlag) and Volker Nannen. "A short introduction to Model Selection, Kolmogorov Complexity and Minimum Description Length", online publication. There is similar diversity in how to use or define multivariate mutual information to measure interactions. See for example reviews and proposals by A. Margolin, K. Wang, A. Califano, and I. Nemenman, Multivariate Dependence and Genetic Networks Inference. IET Syst Biol 4, 428 (2010). A. Jakulin and I. Bratko, Quantifying and Visualizing attribute interactions, (2003b) arxiv.org/abs/cs/0308002v1 online publication. A basic problem is that feature reduction based upon mutual information will not discriminate linear and nonlinear effects for a given variable, but this confusion is compounded with interactions. For example, some multivariate mutual information definitions of interaction like interaction information have surprising negative sign relationships that can defy simple interpretation. In addition, these methods might be interpreted to require main effects be included prior to selecting interactions just like the generalized linear model avoids the marginality scaling problem (J.A. Nelder op cit.), but this can exclude important interaction effects or include superfluous main effects.

3. Rice, D.M. *Society for Neuroscience Abstracts*, 2013, op. cit.
4. Davies, P., Keynote address, American Geriatric Society, 1990.
5. http://archive.org/details/KQED_20120206_200000_Charlie_Rose. The idea that Alzheimer's could start in the medial temporal lobe was also proposed on the basis of evidence that damage at the end was substantially heavier in entorhinal cortex, but there was still not understanding from such neuropathology data of how the onset might occur at least 10 years earlier than the clinical diagnosis. See G.W. Van Hoesen, B.T. Hyman, and A.R. Damasio, Entorhinal Cortex Pathology in Alzheimer's Disease, *Hippocampus* 1(1) (1991):1–8. Our work did provide a direct linkage between recent memory and temporal lobe abnormalities in living older people who were not demented and a very precise estimate of the 10 year average time delay after onset of these abnormalities until clinical diagnosis of dementia.
6. Kahneman, D. Thinking, Fast and Slow, op. cit.
7. Ibid.
8. Ibid.
9. Ibid.
10. Ibid.
11. Ibid.
12. Ibid.
13. Shi, L. et al., "The MicroArray Quality Control (MAQC)-II Study of Common Practices for the Development and Validation of Microarray-Based Predictive Models", op. cit.
14. http://www.pbs.org/wgbh/nova/body/cracking-your-genetic-code.html, March 29, 2012.
15. This website link, http://message.snopes.com/showthread.php?t=30482, has a discussion about this ad back in 2008 amongst women car buyers with a link to the you tube video.
16. Lowenstein, R. *When Genius Failed: the Rise and Fall of Long-Term Capital Management*, (2000, Random House Trade Paperbacks, New York).
17. Taleb, N.N. *The Black Swan: The Impact of the Highly Improbable*, (2007, Random House, New York).

18. Salmon, F. "The Formula that Killed Wall Street", *Wired Magazine* February 23, 2009.
19. Abrahams, C.R. and Zhang, M. Credit Risk Assessment: The New Lending System for Borrowers, Lenders, and Investors (2009, Wiley and SAS Business Series, Hoboken, N.J. and Cary, N.C.).
20. Morgenson, G. and Story, L. "Banks Collapse in Europe Points to Global Risks", *New York Times* October 22, 2011.
21. Derman, E. Models Behaving Badly, op. cit.
22. Tierney, J. "Findings; Social Scientist Sees Bias Within", *New York Times* February 8, 2011.
23. Krugman, P. "The Excel Depression", Opinion Editorial *New York Times* April 19, 2013.

Appendix

1. Golan, A, Judge, G. and Perloff, J.M., op. cit.
2. Mount, J., op. cit.
3. Luce, R.D. and Suppes, P., op. cit.; McFadden, D., op. cit.
4. Hosmer, D.W. and Lemeshow, S., op. cit.
5. Mount, J., op. cit.
6. Mitchell, T., op. cit.
7. Hopefully, there is no confusion here in using t to reflect time with the Student's t or other t distribution values used to compute the error model.
8. Allison, P.D. *Sociological Methodology*, op. cit.
9. Hedeker, D., op. cit. He reviews how other parameters can be temporal dependent. Although as stated in Chapter 4, RELR does not require multilevel parameters so his comments about multilevel parameters do not apply.
10. King, G. and Langche, Z., op. cit.
11. Topsøe, F., op. cit.

INDEX

Note: Page numbers with "'*f*'" denote figures; "*t*" tables.

Printed and bound by CPI Group (UK) Ltd, Croydon, CR0 4YY

03/10/2024

01040425-0008